The Ideas of Algebra, K–12

1988 Yearbook

Arthur F. Coxford
1988 Yearbook Editor
University of Michigan

Albert P. Shulte
General Yearbook Editor
Oakland Schools, Pontiac, Michigan

National Council of
Teachers of Mathematics

Printed in the United States of America

Contents

iii

PART 3: EQUATIONS AND EXPRESSIONS IN ALGEBRA

PART 4: PROBLEM SOLVING IN ALGEBRA

PART 5: USING COMPUTERS AND CALCULATORS TO LEARN ALGEBRA

PART 6: ALGEBRA TEACHING IDEAS

Preface

The publication of the 1985 Yearbook, *The Secondary School Mathematics Curriculum,* by the National Council of Teachers of Mathematics ended a long period during which the subject matter curriculum in secondary mathematics was de-emphasized while we attended to other issues such as problem solving, estimation, the use of calculators and computers, the shortage of mathematics teachers, and the upheaval created by several national reports critical of the educational system. That yearbook paved the way for the present work. It suggested contrasting forces in school algebra. Algebra is a relatively stable curricular area—one in which change is difficult to accomplish in spite of the realization that fundamental ideas are not being well understood by learners. School algebra is the area in which symbolic conventions needed for more advanced mathematics are first developed, yet the main emphasis is often manipulation.

Even though a work on the secondary curriculum gave rise to this work, the NCTM Educational Materials Committee wisely expanded its purview to include the entire curriculum. Thus the editorial panel began its work with the charge to produce a work on algebraic concepts in the K–12 curriculum. A call for papers stimulated nearly 85 submissions. Fewer than half of these could be accepted owing to the space limitations imposed on NCTM yearbooks.

This volume is organized into six parts, as indicated in the Contents. Chapters 1–5, which make up Part 1, first discuss the forces impinging on algebra in the curriculum and suggest possible directions for change. Variables are central, and the several ways they are used in algebra highlight the complexity of the concept and may partially explain the difficulty experienced by the learner. Specifics concerning students' difficulties with variables as well as with function concepts are discussed at some length. Some of the observed difficulties may result from widely accepted instructional practices in grades K–6. These chapters clearly indicate that we need to think carefully about teaching algebra; it is not as simple as one might think!

Chapters 6–8, Part 2, concentrate on concepts and teaching possibilities available prior to the formal introduction of algebra. They reinforce the notion that algebraic ideas are accessible to younger learners but, again, that care is needed in their development. In chapters 9–11, Part 3, equations and expressions in algebra are the focus. It is suggested that previous learning

vii

affects the learner's view of new ideas, that the solution of linear equations demands a careful developmental sequence, and that factoring and polynomials are important algebraic topics. Part 4, made up of chapters 12–14, emphasizes suggestions for teaching and using word problems in algebra.

Part 5, chapters 15–24, emphasizes the use of technology in the algebra classroom. The suggestions include presenting guidelines for the development of instructional software, using computer programming as a way to introduce mathematical concepts, using computer-generated data—tables, graphs, and spreadsheets—to teach concepts and processes, and using the calculator as a tool to introduce the concept of logarithm. The message is clear: Technology is a valuable ally in algebraic instruction.

The yearbook concludes with Part 6, chapters 25–34, whose focus is on helpful, teacher-tested ideas for teaching algebra. These suggestions will find their way into classrooms and make them more productive places for student and teacher.

The task of developing guidelines for the direction and contents of the 1988 Yearbook, reviewing and selecting manuscripts, and making suggestions for improving manuscripts was the task of the editorial panel. It was made up of the general editor, the issue editor, and four individuals of extraordinary experience and expertise in school algebra. My sincere thanks and appreciation are extended to each of these panel members:

Martin Cohen	University of Pittsburgh, Pittsburgh, Pennsylvania
Patricia Fraze	Huron High School, Ann Arbor, Michigan
Peggy House	University of Minnesota, Minneapolis, Minnesota
Edward Rathmell	University of Northern Iowa, Cedar Falls, Iowa
Al Shulte	Oakland Schools, Pontiac, Michigan

A special note of gratitude is owed to Al Shulte, general yearbook editor, who fully participated in all deliberations and who always provided valuable procedural and substantive advice. The editor is also deeply grateful to Charles Hucka and the able production staff of the NCTM Reston office for their editorial contributions. Finally, and most important, deep appreciation is expressed to all the professionals who submitted manuscripts for this yearbook. If it becomes a useful addition to the literature, it is because of their talent and dedication to completing an important task.

The message of this yearbook is that algebra is an exciting, living, and growing school mathematical field. We have had some success in teaching algebraic concepts, but also some failures. We need to continue to learn and to improve our presentation of this subject so that every youngster will experience the thrill of understanding.

ARTHUR F. COXFORD
1988 Yearbook Editor

Part 1
Algebra: Ideas and Issues

1

Reshaping School Algebra: Why and How?

Peggy A. House

A LGEBRA," wrote an accelerated seventh-grade pupil, "is quite hard, and although very educational, it is very frustrating ninety percent of the time. It means hours of instruction that you don't even come close to understanding."

Added a classmate: "I don't know much about algebra, but who cares?"

Who, indeed? And why a yearbook on the topic of algebra in the school curriculum? Why now?

Algebra has long enjoyed a place of distinction in the mathematics curriculum, representing for many students both the culmination of years of study in arithmetic and the beginning of more years of study in other branches of mathematics. Few have contested its importance, although many, like the pupils quoted above, have only shallow notions of its meaning and significance.

Today's realities make it imperative that we examine anew the full range of the mathematics curriculum and the manner in which it is taught. Recent NCTM yearbooks have approached this task by examining the curriculum in toto (1985) and the geometry curriculum in particular (1987). This yearbook is a natural extension of that effort.

Periodic review of the curriculum and of related instructional practices is not a new idea. Most mathematics educators active today have witnessed several swings of the curricular pendulum with the rise and fall of movements like the "new math" and "back to the basics." What sets this round of curricular examination apart from previous ones is not so much a function of mathematics itself as a function of cultural changes in the world around us.

At center stage in this scenario of rapid change is computer technology. It is unnecessary to chronicle the extent to which computers have already transformed the way we live, the way we work, the way we spend our leisure time. For persons not on the frontier of technological developments, these changes can be so remarkable that many can neither comprehend the impact of present technology nor imagine the magnitude and scope of changes yet to come.

Yet the impact of computer technology on how and what students learn and on how and what teachers teach has lagged behind its influence on manufacturing, product development, sales, service industries, and many other spheres of human activity. In many classrooms students continue to be drilled on the possession of information and the development of competence in the performance of algorithmic manipulations. And although appropriate levels of factual knowledge and skill are important outcomes of the algebra program, what students need even more is a sound understanding of algebraic concepts and the ability to use knowledge in new and often unexpected ways.

We have customarily looked to the records of the past for insight into making decisions for the future. Often the future we have projected has been largely a rearrangement of elements from the past. We have incorporated trigonometry into intermediate algebra, integrated solid with plane geometry, permuted the order of presentation and emphasis given to various topics, and introduced some new concepts with varying degrees of permanence. But for all that, the content of school mathematics in 1988 still bears a striking resemblance to the school mathematics of 1928, of 1948, or of 1968.

Now we are asked to contemplate changes qualitatively different from those we have encountered within the working memory of active mathematics educators. Recommendations have been made suggesting that we deemphasize or eliminate topics dear to the hearts of many teachers, such as trinomial factoring or algebraic fractions. New delivery systems are proposed that incorporate programming, spreadsheets, and symbol manipulators to alter not only how we teach but what we teach. A reordering of priorities among learning outcomes, the addition of new content such as discrete mathematics, and the requirement of more mathematics for most students are among the common themes of school mathematics reform. And throughout these various scenarios of probable and desirable futures, the role of algebra is, if anything, enhanced and strengthened.

FORCES ON ALGEBRA IN SCHOOL MATHEMATICS

The forces operating on the content, instruction, and use of algebra arise primarily from two sources. Each force has the potential to alter the way algebra is taught in school, and each warrants careful examination and evaluation.

Computing Technology

A major force impinging on the curriculum is computing technology, which includes both computers and calculators. Surprisingly, perhaps, educators have thus far been more accepting of computers than of calculators. Yet even a simple four-function calculator marks topics like computation with logarithms and the calculation of square roots for almost certain elimination from the syllabus, even while the concepts of logarithmic and exponential functions grow in importance. Further, when cumbersome computations are accomplished with a calculator, problems can be made more realistic and more complex. And the ability to evaluate functions and expressions quickly and accurately allows the teacher to de-emphasize some analytical processes, such as simplifying algebraic expressions, and to stress instead such heuristics as making tables and looking for patterns.

But even relatively modest changes engender resistance on the part of some educators who may not be able to imagine a curriculum devoid of traditional topics and not based on traditional teaching methods. Educators tend to forget that once, before the development of paper and printing technology, computation was accomplished by manipulating physical counters. Paper and printing technology changed that. Algorithms were developed that replaced object manipulation with symbol manipulation, thus enabling persons to perform more complex mathematics, which ultimately led to significant new developments in both mathematics and related disciplines, especially science. How much of modern algebra, analysis, trigonometry, calculus, and other branches of mathematics depend on that one technological breakthrough?

The impact of computing technology will be even more profound. Its influence will be felt not only as a force for change but also as a goal of curricular change and as a means for instructional development. In algebra, as in other branches of mathematics, this force already is apparent.

With the advent of computing technology, disciplines other than the physical and engineering sciences, such as business and the social and biological sciences, have become highly dependent on mathematical processes. But the mathematics central for them is not the continuous mathematics of analysis and the calculus but discrete mathematics, which includes probability, statistics, matrix algebra, and finite systems. In these areas algebraic concepts and processes such as manipulating variables and evaluating trends are of fundamental importance.

The traditional school mathematics curriculum has offered only two options to the mathematics student: a "practical" or consumer-oriented general mathematics option, frequently entered into with little or no algebra; and an academic, college-preparatory strand that is calculus preparatory. Neither sequence has included many topics of discrete mathematics. Thus the social, management, and biological disciplines are calling for an academic alternative to the traditional path to calculus. Such alternatives would emphasize the discrete mathematics for which algebra provides an essential foundation and would stress its importance to a much larger proportion of the student and adult populations than has been true in the past. The demands for such alternatives can be expected to continue.

At the same time, computing technology will have implications for what we teach within the algebra curriculum. Somewhat paradoxically, the place of algorithms will be both diminished and enhanced—diminished in the area of memorizing algorithms for the purpose of turning out answers, but enhanced in the direction of learning to plan and design algorithms for human and computer execution.

The recent debate over the place of such mainstays of the algebra curriculum as trinomial factoring and the simplification of rational expressions brings to light other concerns. Factoring, for example, traditionally is taught as a means to finding roots of polynomial functions. What are the implications for factoring skills when high-speed computing technology can use an iterative approach to approximate zeros of functions with successively greater degrees of accuracy and can do so in less time than the student can factor and evaluate written expressions? In addition, there exists a class of software known as symbol manipulators that can perform not only numeric evaluations of algebraic expressions but also symbolic manipulation of mathematical expressions. Such software can be expected to continue to improve in capability and to be available in hand-held packages. Will such programs relegate factoring and other topics to the same status as computation with Napier's rods or the abacus?

For more than a decade now mathematics educators have argued for a definition of basic mathematical skills that extends beyond computation. In algebra, too, basic algebraic skill must be conceived as encompassing more than symbol manipulation. Of fundamental significance are an understanding of concepts such as variable and function; the representation of phenomena in algebraic and graphical form; and facility in the presentation and interpretation of data, estimation and approximation, prediction, and problem formulation as well as problem solving. The new basics of algebra may be expected to include as well the manipulation and interpretation of spreadsheets and processes such as "sort," "locate," "insert," or "merge."

The impact of changes such as those suggested in this section will be felt not only in the secondary school mathematics class but in the elementary and middle school prealgebra curriculum as well. Students who design flow charts or program an algorithm, who collect and organize tables of data, who evaluate variable expressions with a computer or calculator, or who ask "What if?" questions with a spreadsheet are laying an important foundation for the formal study of algebra. At all levels the influence of technology will be felt on both the curriculum and the instructional process.

It is not surprising that early applications of technology to instruction frequently modeled the behaviors that teachers employ most frequently: direct teaching and supervising student work. Much of the first software available for mathematics was either tutorial or drill-and-practice in nature—the former providing the electronic counterpart of direct teaching, the latter serving as a monitor of electronic seatwork. The only really notable exception to these two uses was in programming, where the computer itself became the object, not the medium, of instruction.

Now, however, we have available microcomputer software that can and should have a significant influence on classroom instructional practices. Graphing packages, for example, do what no chalkboard or overhead projector can do, and they provide the teacher with dynamic means of demonstrating and exploring important concepts such as the behavior of functions and their graphs. Electronic spreadsheets enable the teacher and student to conduct "What if?" investigations, such as "What if you change the argument of the function?" or "What if you change the hypothesis to . . . ?" Answers are delivered almost instantaneously, and learning can focus on conceptual development unfettered by cumbersome computations. Simulations of "real world" problems made impractical by paper-and-pencil methods can now become commonplace.

The learning environment suggested by these developments in technology bears little resemblance to the traditional algebra classroom, and history reminds us that educational changes do not come easily. In addition, forces other than technology impinge on the algebra curriculum and its instruction. These forces may or may not facilitate improvements in the way the subject is taught and learned, but they, along with the technological changes, are forces with which we must reckon.

Social Forces

As suggested earlier, the impact of technology on virtually all phases of the workplace has created new demands for citizens with facility in quantitative reasoning and mathematical processes. This includes a knowledge of a range of topics, such as statistics and probability, that are not yet in the common curriculum for the average school pupil. Likewise, in the wake of several national reports critical of American schooling, secondary school graduation requirements and college and university entrance standards have begun to move in the direction of more mathematics required of more students. One result will be a larger, more heterogeneous enrollment in classes at the level of algebra and above, which, in turn, has implications for both teachers and curriculum developers, who must accommodate a wider range of interests and abilities.

At the same time, data from national and international assessments of educational achievement have made it clear that the achievement of American pupils on all but the lowest levels of mathematical knowledge and algorithmic manipulation is well below what educators and the public consider acceptable. To reverse this trend will require increased human, financial, and curricular resources. Unfortunately, these resources do not seem to be forthcoming in most school districts and states.

Yet the public continues to demand accountability and demonstrated excellence, which too often is interpreted to mean higher scores on standardized tests. On the one hand, there is an urgent need for new approaches to evaluation that adequately assess higher-order thinking and problem-solving objectives. On the other hand, there is an ever-present danger of responding to demands for higher scores either by weakening the assessment process even further or by "teaching to the test," emphasizing only those mechanical skills that are easily measured.

We are faced as well with a crisis in the mathematics teaching community. Severe shortages of mathematics teachers have become a national problem in recent years. The pending retirements of a large segment of the active teaching cadre within the next few years will exacerbate the situation. In many school systems the impact of the teacher shortage is felt most keenly in the junior high or middle school, where minimally qualified or out-of-field teachers are recruited for mathematics instruction and those who are prepared in mathematics are assigned to classes at the senior high school level. Algebra, a common junior high school subject, is vulnerable to the consequences of such decisions; yet algebra often marks a turning point in a pupil's choice to continue or not in the study of mathematics. The quality of instruction in that course can be a critical variable in pupils' decisions. What is more, teachers with minimal qualifications or experience in mathematics are unlikely to effect the curricular and instructional changes that are needed in algebra at this time.

FINAL COMMENT

Teachers seek answers to questions about teaching effectiveness and information on new methods and materials of instruction. Curriculum developers require insights into contemporary issues and needs as well as solid foundations in mathematics and the psychology of learning. Mathematics educators at all levels benefit when presented with a forum for the discussion of ideas and issues. It is the goal of this yearbook to contribute to that forum by presenting information and perspectives on the teaching of algebraic concepts in the K–12 curriculum.

The task of modifying the algebra curriculum in response to the forces pulling it in diverse directions is a significant one that will not be accomplished without major, sometimes painful, efforts. And although definitive answers to educational questions may perhaps remain forever beyond our reach, it is only through honest assessment and open exchange that we can hope to improve the conditions of teaching necessary to demonstrate to our students that someone does, indeed, care.

REFERENCES

National Council of Teachers of Mathematics. *Learning and Teaching Geometry, K–12*. 1987 Yearbook, edited by Mary Montgomery Lindquist. Reston, Va.: The Council, 1987.

———. *The Secondary School Mathematics Curriculum*. 1985 Yearbook, edited by Christian R. Hirsch. Reston, Va.: The Council, 1985.

CAN YOUR ALGEBRA CLASS SOLVE THIS?*

Problem 1. The graphs of $2y + 3 + x = 0$ and $3y + ax + 2 = 0$ are lines. If the lines are perpendicular, find the value of a.

Solution on page 248

CAN YOUR ALGEBRA CLASS SOLVE THIS?

Problem 2. What is the value of c if the vertex of the parabola $y = x^2 - 8x + c$ is a point on the x-axis?

Solution on page 248

---

*This is the first of twenty-four problems you will find interspersed among the articles of the book. They are supplied for use in the algebra classroom by Terry Goodman, Central Missouri State University, Warrensburg, and Martin Cohen, University of Pittsburgh. Solutions appear on page 248.

2

Conceptions of School Algebra and Uses of Variables

Zalman Usiskin

WHAT IS SCHOOL ALGEBRA?

A LGEBRA is not easily defined. The algebra taught in school has quite a different cast from the algebra taught to mathematics majors. Two mathematicians whose writings have greatly influenced algebra instruction at the college level, Saunders Mac Lane and Garrett Birkhoff (1967), begin their *Algebra* with an attempt to bridge school and university algebras:

> Algebra starts as the art of manipulating sums, products, and powers of numbers. The rules for these manipulations hold for all numbers, so the manipulations may be carried out with letters standing for the numbers. It then appears that the same rules hold for various different sorts of numbers . . . and that the rules even apply to things . . . which are not numbers at all. An algebraic system, as we will study it, is thus a set of elements of any sort on which functions such as addition and multiplication operate, provided only that these operations satisfy certain basic rules. (P. 1)

If the first sentence in the quote above is thought of as arithmetic, then the second sentence is school algebra. For the purposes of this article, then, school algebra has to do with the understanding of "letters" (today we usually call them *variables*) and their operations, and we consider students to be studying algebra when they first encounter variables.

However, since the concept of variable itself is multifaceted, reducing algebra to the study of variables does not answer the question "What is school algebra?" Consider these equations, all of which have the same form—the product of two numbers equals a third:

8

1. $A = LW$
2. $40 = 5x$
3. $\sin x = \cos x \cdot \tan x$
4. $1 = n \cdot (1/n)$
5. $y = kx$

Each of these has a different feel. We usually call (1) a formula, (2) an equation (or open sentence) to solve, (3) an identity, (4) a property, and (5) an equation of a function of direct variation (not to be solved). These different names reflect different uses to which the idea of variable is put. In (1), A, L, and W stand for the quantities area, length, and width and have the feel of knowns. In (2), we tend to think of x as unknown. In (3), x is an argument of a function. Equation (4), unlike the others, generalizes an arithmetic pattern, and n identifies an instance of the pattern. In (5), x is again an argument of a function, y the value, and k a constant (or parameter, depending on how it is used). Only with (5) is there the feel of "variability," from which the term *variable* arose. Even so, no such feel is present if we think of that equation as representing the line with slope k containing the origin.

Conceptions of variable change over time. In a text of the 1950s (Hart 1951a), the word *variable* is not mentioned until the discussion of systems (p. 168), and then it is described as "a changing number." The introduction of what we today call variables comes much earlier (p. 11), through formulas, with these cryptic statements: "In each formula, the letters represent numbers. *Use of letters to represent numbers is a principal characteristic of algebra*" (Hart's italics). In the second book in that series (Hart 1951b), there is a more formal definition of variable (p. 91): "A variable is a literal number that may have two or more values during a particular discussion."

Modern texts in the late part of that decade had a different conception, represented by this quote from May and Van Engen (1959) as part of a careful analysis of this term:

> Roughly speaking, a variable is a symbol for which one substitutes names for some objects, usually a number in algebra. A variable is always associated with a set of objects whose names can be substituted for it. These objects are called values of the variable. (P. 70)

Today the tendency is to avoid the "name-object" distinction and to think of a variable simply as a symbol for which things (most accurately, things from a particular replacement set) can be substituted.

The "symbol for an element of a replacement set" conception of variable seems so natural today that it is seldom questioned. However, it is not the only view possible for variables. In the early part of this century, the formalist school of mathematics considered variables and all other mathematics symbols merely as marks on paper related to each other by assumed or derived properties that are also marks on paper (Kramer 1981).

Although we might consider such a view tenable to philosophers but impractical to users of mathematics, present-day computer algebras such as MACSYMA and muMath (see Pavelle, Rothstein, and Fitch [1981]) deal with letters without any need to refer to numerical values. That is, today's computers can operate as both experienced and inexperienced users of algebra do operate, blindly manipulating variables without any concern for, or knowledge of, what they represent.

Many students think all variables are letters that stand for numbers. Yet the values a variable takes are not always numbers, even in high school mathematics. In geometry, variables often represent points, as seen by the use of the variables A, B, and C when we write "if $AB = BC$, then$\triangle ABC$ is isosceles." In logic, the variables p and q often stand for propositions; in analysis, the variable f often stands for a function; in linear algebra the variable A may stand for a matrix, or the variable v for a vector, and in higher algebra the variable $*$ may represent an operation. The last of these demonstrates that variables need not be represented by letters.

Students also tend to believe that a variable is always a letter. This view is supported by many educators, for

$$3 + x = 7 \text{ and } 3 + \triangle = 7$$

are usually considered algebra, whereas

$$3 + \underline{\hspace{1cm}} = 7 \text{ and } 3 + ? = 7$$

are not, even though the blank and the question mark are, in this context of desiring a solution to an equation, logically equivalent to the x and the $\triangle$.

In summary, variables have many possible definitions, referents, and symbols. Trying to fit the idea of variable into a single conception oversimplifies the idea and in turn distorts the purposes of algebra.

TWO FUNDAMENTAL ISSUES IN ALGEBRA INSTRUCTION

Perhaps the major issue surrounding the teaching of algebra in schools today regards the extent to which students should be required to be able to do various manipulative skills by hand. (Everyone seems to acknowledge the importance of students having *some* way of doing the skills.) A 1977 NCTM-MAA report detailing what students need to learn in high school mathematics emphasizes the importance of learning and practicing these skills. Yet more recent reports convey a different tone:

> The basic thrust in Algebra I and II has been to give students moderate technical facility. . . . In the future, students (and adults) may not have to do much algebraic manipulation. . . . Some blocks of traditional drill can surely be curtailed. (CBMS 1983, p. 4)

A second issue relating to the algebra curriculum is the question of the role of functions and the timing of their introduction. Currently, functions are treated in most first-year algebra books as a relatively insignificant topic and first become a major topic in advanced or second-year algebra. Yet in some elementary school curricula (e.g., CSMP 1975) function ideas have been introduced as early as first grade, and others have argued that functions should be used as the major vehicle through which variables and algebra are introduced.

It is clear that these two issues relate to the very purposes for teaching and learning algebra, to the goals of algebra instruction, to the conceptions we have of this body of subject matter. What is not as obvious is that they relate to the ways in which variables are used. In this paper I try to present a framework for considering these and other issues relating to the teaching of algebra. My thesis is that the purposes we have for teaching algebra, the conceptions we have of the subject, and the uses of variables are inextricably related. *Purposes for algebra* are determined by, or are related to, different *conceptions of algebra,* which correlate with the different relative importance given to various *uses of variables.*

Conception 1: Algebra as generalized arithmetic

In this conception, it is natural to think of variables as pattern generalizers. For instance, $3 + 5.7 = 5.7 + 3$ is generalized as $a + b = b + a$. The pattern

$$3 \cdot 5 = 15$$
$$2 \cdot 5 = 10$$
$$1 \cdot 5 = 5$$
$$0 \cdot 5 = 0$$

is extended to give multiplication by negatives (which, in this conception, is often considered algebra, not arithmetic):

$$-1 \cdot 5 = -5$$
$$-2 \cdot 5 = -10$$

This idea is generalized to give properties such as

$$-x \cdot y = -xy.$$

At a more advanced level, the notion of variable as pattern generalizer is fundamental in mathematical modeling. We often find relations between numbers that we wish to describe mathematically, and variables are exceedingly useful tools in that description. For instance, the world record T (in seconds) for the mile run in the year Y since 1900 is rather closely described by the equation

$$T = -0.4Y + 1020.$$

This equation merely generalizes the arithmetic values found in many almanacs. In 1974, when the record was 3 minutes 51.1 seconds and had not changed in seven years, I used this equation to predict that in 1985 the record would be 3 minutes 46 seconds (for graphs, see Usiskin [1976] or Bushaw et al. [1980]). The actual record at the end of 1985 was 3 minutes 46.31 seconds.

The key instructions for the student in this conception of algebra are *translate* and *generalize*. These are important skills not only for algebra but also for arithmetic. In a compendium of applications of arithmetic (Usiskin and Bell 1984), Max Bell and I concluded that it is impossible to adequately study arithmetic without implicitly or explicitly dealing with variables. Which is easier, "The product of any number and zero is zero" or "For all n, $n \cdot 0 = 0$"? The superiority of algebraic over English language descriptions of number relationships is due to the similarity of the two syntaxes. The algebraic description looks like the numerical description; the English description does not. A reader in doubt of the value of variables should try to describe the rule for multiplying fractions first in English, then in algebra.

Historically, the invention of algebraic notation in 1564 by François Viète (1969) had immediate effects. Within fifty years, analytic geometry had been invented and brought to an advanced form. Within a hundred years, there was calculus. Such is the power of algebra as generalized arithmetic.

Conception 2: Algebra as a study of procedures for solving certain kinds of problems

Consider the following problem:

When 3 is added to 5 times a certain number, the sum is 40. Find the number.

The problem is easily translated into the language of algebra:

$$5x + 3 = 40$$

Under the conception of algebra as a generalizer of patterns, we do not have unknowns. We generalize known relationships among numbers, and so we do not have even the feeling of unknowns. Under that conception, this problem is finished; we have found the general pattern. However, under the conception of algebra as a study of procedures, we have only begun.

We solve with a procedure. Perhaps add -3 to each side:

$$5x + 3 + -3 = 40 + -3$$

Then simplify (the number of steps required depends on the level of student and preference of the teacher):

$$5x = 37$$

Now solve this equation in some way, arriving at $x = 7.4$. The "certain number" in the problem is 7.4, and the result is easily checked.

In solving these kinds of problems, many students have difficulty moving from arithmetic to algebra. Whereas the arithmetic solution ("in your head") involves subtracting 3 and dividing by 5, the algebraic form $5x + 3$ involves multiplication by 5 and addition of 3, the inverse operations. That is, to set up the equation, you must think precisely the opposite of the way you would solve it using arithmetic.

In this conception of algebra, variables are either *unknowns* or *constants*. Whereas the key instructions in the use of a variable as a pattern generalizer are translate and generalize, the key instructions in this use are *simplify* and *solve*. In fact, "simplify" and "solve" are sometimes two different names for the same idea: For example, we ask students to solve $|x - 2| = 5$ to get the answer $x = 7$ or $x = -3$. But we could ask students, "Rewrite $|x - 2| = 5$ without using absolute value." We might then get the answer $(x - 2)^2 = 25$, which is another equivalent sentence.

Polya (1957) wrote, "If you cannot solve the proposed problem try to solve first some related problem" (p. 31). We follow that dictum literally in solving most sentences, finding equivalent sentences with the same solution. We also simplify expressions so that they can more easily be understood and used. To repeat: simplifying and solving are more similar than they are usually made out to be.

Conception 3: Algebra as the study of relationships among quantities

When we write $A = LW$, the area formula for a rectangle, we are describing a relationship among three quantities. There is not the feel of an unknown, because we are not solving for anything. The feel of formulas such as $A = LW$ is different from the feel of generalizations such as $1 = n \cdot (1/n)$, even though we can think of a formula as a special type of generalization.

Whereas the conception of algebra as the study of relationships may begin with formulas, the crucial distinction between this and the previous conceptions is that, here, variables *vary*. That there is a fundamental difference between the conceptions is evidenced by the usual response of students to the following question:

What happens to the value of $1/x$ as x gets larger and larger?

The question seems simple, but it is enough to baffle most students. We have not asked for a value of x, so x is not an unknown. We have not asked the student to translate. There is a pattern to generalize, but it is not a pattern that looks like arithmetic. (It is not appropriate to ask what happens to the value of 1/2 as 2 gets larger and larger!) It is fundamentally an algebraic pattern. Perhaps because of its intrinsic algebraic nature, some mathematics educators believe that algebra should first be introduced through this use of variable. For instance, Fey and Good (1985) see the following as the key questions on which to base the study of algebra:

> For a given function $f(x)$, find—
> 1. $f(x)$ for $x = a$;
> 2. x so that $f(x) = a$;
> 3. x so that maximum or minimum values of $f(x)$ occur;
> 4. the rate of change in f near $x = a$;
> 5. the average value of f over the interval (a,b). (P. 48)

Under this conception, a variable is an *argument* (i.e., stands for a domain value of a function) or a *parameter* (i.e., stands for a number on which other numbers depend). Only in this conception do the notions of independent variable and dependent variable exist. Functions arise rather immediately, for we need to have a name for values that depend on the argument or parameter x. Function notation (as in $f(x) = 3x + 5$) is a new idea when students first see it: $f(x) = 3x + 5$ looks and feels different from $y = 3x + 5$. (In this regard, one reason $y = f(x)$ may confuse students is because the function f, rather than the argument x, has become the parameter. Indeed, the use of $f(x)$ to name a function, as Fey and Good do in the quote above, is seen by some educators as contributing to that confusion.)

That variables as arguments differ from variables as unknowns is further evidenced by the following question:

Find an equation for the line through (6,2) with slope 11.

The usual solution combines all the uses of variables discussed so far, perhaps explaining why some students have difficulty with it. Let us analyze the usual solution. We begin by noting that points on a line are related by an equation of the form

$$y = mx + b.$$

This is both a pattern among variables and a formula. In our minds it is a function with domain variable x and range variable y, but to students it is not clear which of m, x, or b is the argument. As a pattern it is easy to understand, but in the context of this problem, some things are unknown. All the letters look like unknowns (particularly the x and y, letters traditionally used for that purpose).

Now to the solution. Since we know m, we substitute for it:

$$y = 11x + b$$

Thus m is here a constant, not a parameter. Now we need to find b. Thus b has changed from parameter to unknown. But how to find b? We use one pair of the many pairs in the relationship between x and y. That is, we select a value for the argument x for which we know y. Having to substitute a pair of values for x and y can be done because $y = mx + b$ describes a general pattern among numbers. With substitution,

$$2 = 11 \cdot 6 + b,$$

and so $b = -64$. But we haven't found x and y even though we have values for them, because they were not unknowns. We have only found the unknown b, and we substitute in the appropriate equation to get the answer

$$y = 11x - 64.$$

Another way to make the distinction between the different uses of the variables in this problem is to engage quantifiers. We think: For all x and y, there exist m and b with $y = mx + b$. We are given the value that exists for m, so we find the value that exists for b by using one of the "for all x and y" pairs, and so on. Or we use the equivalent set language: We know the line is $\{(x,y): y = mx + b\}$ and we know m and try to find b. In the language of sets or quantifiers, x and y are known as *dummy variables* because any symbols could be used in their stead. It is rather hard to convince students and even some teachers that $\{x: 3x = 6\} = \{y: 3y = 6\}$, even though each set is $\{2\}$. Many people think that the function f with $f(x) = x + 1$ is not the same as the function g with the same domain as f and with $g(y) = y + 1$. Only when variables are used as arguments may they be considered as dummy variables; this special use tends to be not well understood by students.

Conception 4: Algebra as the study of structures

The study of algebra at the college level involves structures such as groups, rings, integral domains, fields, and vector spaces. It seems to bear little resemblance to the study of algebra at the high school level, although the fields of real numbers and complex numbers and the various rings of polynomials underlie the theory of algebra, and properties of integral domains and groups explain why certain equations can be solved and others not. Yet we recognize algebra as the study of structures by the properties we ascribe to operations on real numbers and polynomials. Consider the following problem:

$$\text{Factor } 3x^2 + 4ax - 132a^2.$$

The conception of variable represented here is not the same as any previously discussed. There is no function or relation; the variable is not an argument. There is no equation to be solved, so the variable is not acting as an unknown. There is no arithmetic pattern to generalize.

The answer to the factoring question is $(3x + 22a)(x - 6a)$. The answer could be checked by substituting values for x and a in the given polynomial and in the factored answer, but this is almost never done. If factoring were checked that way, there would be a wisp of an argument that here we are generalizing arithmetic. But in fact, the student is usually asked to check by multiplying the binomials, exactly the same procedure that the student has employed to get the answer in the first place. It is silly to check by repeating the process used to get the answer in the first place, but in this kind of problem students tend to treat the variables as marks on paper, without numbers as a referent. In the conception of algebra as the study of structures, the variable is little more than an arbitrary symbol.

There is a subtle quandary here. We want students to have the referents (usually real numbers) for variables in mind as they use the variables. But we also want students to be able to operate on the variables without always having to go to the level of the referent. For instance, when we ask students to derive a trigonometric identity such as $2\sin^2 x - 1 = \sin^4 x - \cos^4 x$, we do not want the student to think of the sine or cosine of a specific number or even to think of the sine or cosine functions, and we are not interested in ratios in triangles. We merely want to manipulate $\sin x$ and $\cos x$ into a different form using properties that are just as abstract as the identity we wish to derive.

In these kinds of problems, faith is placed in properties of the variables, in relationships between x's and y's and n's, be they addends, factors, bases, or exponents. The variable has become an arbitrary object in a structure related by certain properties. It is the view of variable found in abstract algebra.

Much criticism has been leveled against the practice by which "symbol pushing" dominates early experiences with algebra. We call it "blind" manipulation when we criticize; "automatic" skills when we praise. Ultimately everyone desires that students have enough facility with algebraic symbols to deal with the appropriate skills abstractly. The key question is, What constitutes "enough facility"?

It is ironic that the two manifestations of this use of variable—theory and manipulation—are often viewed as opposite camps in the setting of policy toward the algebra curriculum, those who favor manipulation on one side, those who favor theory on the other. They come from the same view of variable.

VARIABLES IN COMPUTER SCIENCE

Algebra has a slightly different cast in computer science from what it has in mathematics. There is often a different syntax. Whereas in ordinary algebra, $x = x + 2$ suggests an equation with no solution, in BASIC the same sentence conveys the replacement of a particular storage location in a computer by a number two greater. This use of variable has been identified by Davis, Jockusch, and McKnight (1978, p. 33):

Computers give us another view of the basic mathematical concept of variable. From a computer point of view, the name of a variable can be thought of as the address of some specific memory register, and the value of the variable can be thought of as the contents of this memory register.

In computer science, variables are often identified strings of letters and numbers. This conveys a different feel and is the natural result of a different setting for variable. Computer applications tend to involve large numbers of variables that may stand for many different kinds of objects. Also, computers are programmed to manipulate the variables, so we do not have to abbreviate them for the purpose of easing the task of blind manipulation.

In computer science the uses of variables cover all the uses we have described above for variables. There is still the generalizing of arithmetic. The study of algorithms is a study of procedures. In fact, there are typical algebra questions that lend themselves to algorithmic thinking:

Begin with a number. Add 3 to it. Multiply it by 2. Subtract 11 from the result. . . .

In programming, one learns to consider the variable as an argument far sooner than is customary in algebra. In order to set up arrays, for example, some sort of function notation is needed. And finally, because computers have been programmed to perform manipulations with symbols without any referents for them, computer science has become a vehicle through which many students learn about variables (Papert 1980). Ultimately, because of this influence, it is likely that students will learn the many uses of variables far earlier than they do today.

SUMMARY

The different conceptions of algebra are related to different uses of variables. Here is an oversimplified summary of those relationships:

Conception of algebra	Use of variables
Generalized arithmetic	Pattern generalizers (translate, generalize)
Means to solve certain problems	Unknowns, constants (solve, simplify)
Study of relationships	Arguments, parameters (relate, graph)
Structure	Arbitrary marks on paper (manipulate, justify)

Earlier in this article, two issues concerning instruction in algebra were mentioned. Given the discussion above, it is now possible to interpret these issues as a question of the relative importance to be given at various levels of study to the various conceptions.

For example, consider the question of paper-and-pencil manipulative skills. In the past, one had to have such skills in order to solve problems and in order to study functions and other relations. Today, with computers able to simplify expressions, solve sentences, and graph functions, what to do with manipulative skills becomes a question of the importance of algebra as a structure, as the study of arbitrary marks on paper, as the study of arbitrary relationships among symbols. The prevailing view today seems to be that this should not be the major criterion (and certainly not the only criterion) by which algebra content is determined.

Consider the question of the role of function ideas in the study of algebra. It is again a question of the relative importance of the view of algebra as the study of relationships among quantities, in which the predominant manifestation of variable is as argument, compared to the other roles of algebra: as generalized arithmetic or as providing a means to solve problems.

Thus some of the important issues in the teaching and learning of algebra can be crystallized by casting them in the framework of conceptions of algebra and uses of variables, conceptions that have been changed by the explosion in the uses of mathematics and by the omnipresence of computers. No longer is it worthwhile to categorize algebra solely as generalized arithmetic, for it is much more than that. Algebra remains a vehicle for solving certain problems but it is more than that as well. It provides the means by which to describe and analyze relationships. And it is the key to the characterization and understanding of mathematical structures. Given these assets and the increased mathematization of society, it is no surprise that algebra is today the key area of study in secondary school mathematics and that this preeminence is likely to be with us for a long time.

REFERENCES

Bushaw, Donald, Max Bell, Henry Pollak, Maynard Thompson, and Zalman Usiskin. *A Sourcebook of Applications of School Mathematics*. Reston, Va.: National Council of Teachers of Mathematics, 1980.

Comprehensive School Mathematics Program. *CSMP Overview*. St. Louis: CEMREL, 1975.

Conference Board of the Mathematical Sciences. *The Mathematical Sciences Curriculum K–12: What Is Still Fundamental and What Is Not*. Report to the NSB Commission on Precollege Education in Mathematics, Science, and Technology. Washington, D.C.: CBMS, 1983.

Davis, Robert B., Elizabeth Jockusch, and Curtis McKnight. "Cognitive Processes in Learning Algebra." *Journal of Children's Mathematical Behavior* 2 (Spring 1978): 1–320.

Fey, James T., and Richard A. Good. "Rethinking the Sequence and Priorities of High School Mathematics Curricula." In *The Secondary School Mathematics Curriculum*, 1985 Yearbook of the National Council of Teachers of Mathematics, pp. 43–52. Reston, Va.: NCTM, 1985.

Hart, Walter W. *A First Course in Algebra*. 2d ed. Boston: D. C. Heath & Co., 1951a.

———. *A Second Course in Algebra*. 2d ed., enlarged. Boston: D. C. Heath & Co., 1951b.

Kramer, Edna E. *The Nature and Growth of Modern Mathematics*. Princeton, N.J.: Princeton University Press, 1981.

Mac Lane, Saunders, and Garrett Birkhoff. *Algebra*. New York: Macmillan Co., 1967.

May, Kenneth O., and Henry Van Engen. "Relations and Functions." In *The Growth of Mathematical Ideas, Grades K–12*, Twenty-fourth Yearbook of the National Council of Teachers of Mathematics, pp. 65–110. Washington, D.C.: NCTM, 1959.

National Council of Teachers of Mathematics and the Mathematical Association of America. *Recommendations for the Preparation of High School Students for College Mathematics Courses*. Reston, Va.: NCTM; Washington, D.C.: MAA, 1977.

Papert, Seymour. *Mindstorms: Children, Computers, and Powerful Ideas*. New York: Basic Books, 1980.

Pavelle, Richard, Michael Rothstein, and John Fitch. "Computer Algebra." *Scientific American*, December 1981, pp. 136–52.

Polya, George, *How to Solve It*. 2d ed. Princeton, N.J.: Princeton University Press, 1957.

Usiskin, Zalman. *Algebra through Applications*. Chicago: Department of Education, University of Chicago, 1976.

Usiskin, Zalman, and Max Bell. *Applying Arithmetic*. Preliminary ed. Chicago: Department of Education, University of Chicago, 1984.

Viète, François. "The New Algebra." In *A Source Book on Mathematics, 1200–1800*, edited by D. J. Struik, pp. 74–81. Cambridge, Mass.: Harvard University Press, 1969.

CAN YOUR ALGEBRA CLASS SOLVE THIS?

Problem 3. Find all real values of x that satisfy
$$(x^2 - 5x + 5)^{x^2 - 9x + 20} = 1.$$

Solution on page 248

CAN YOUR ALGEBRA CLASS SOLVE THIS?

Problem 4. If p pencils cost c cents, how many pencils can be purchased for d dollars?

Solution on page 248

3

Children's Difficulties in Beginning Algebra

Lesley R. Booth

A LGEBRA "is a source of considerable confusion and negative attitudes among pupils." So commented a British study of adults' recollections of their experiences in learning school mathematics (University of Bath 1982). It is, of course, a sentiment that might just as easily have been expressed by any mathematics teacher. No doubt many of their students would also agree. One reason for this state of affairs is that students seem to find algebra difficult.

WHY IS ALGEBRA DIFFICULT TO LEARN?

One way of trying to find out what makes algebra difficult is to identify the kinds of errors students commonly make in algebra and then to investigate the reasons for these errors. One research project that took this approach was the algebra strand of the Strategies and Errors in Secondary Mathematics (SESM) project conducted in the United Kingdom from 1980 to 1983 (Booth 1984). The students involved in this research were in grades 8 to 10 (ages thirteen to sixteen) and had been studying algebra within the context of an integrated mathematics program since grade 7. Consequently the younger students had typically covered work such as simplifying algebraic expressions, simple factorization, the solution of simple linear equations, substituting in formulas, and so on. The older students had also usually studied algebraic fractions and indices, quadratic and simultaneous equations, graphing equalities and inequalities, and more complex factorizations and simplifications. Despite differences in age and experience in algebra, similar errors appeared at each grade level. Interviews with the students who were making these errors showed that many of them could be traced to the students' ideas of such aspects as—

20

a) the focus of algebraic activity and the nature of "answers";
b) the use of notation and convention in algebra;
c) the meaning of letters and variables;
d) the kinds of relationships and methods used in arithmetic.

THE FOCUS OF ALGEBRAIC ACTIVITY AND THE NATURE OF ANSWERS

In arithmetic, the focus of activity is the finding of particular numerical answers. In algebra, however, this is not so. In algebra the focus is on the derivation of procedures and relationships and the expression of these in general, simplified form. One reason for deriving such general statements is to use them as "procedural rules" for solving appropriate problems and hence finding numerical answers, but the immediate focus is on the derivation, expression, and manipulation of the general statement itself. Many students do not realize this; they still assume that what is required is a numerical answer, as fourteen-year-old Wendy did in approaching the problem in figure 3.1.

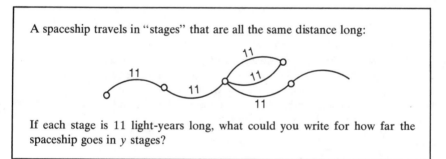

A spaceship travels in "stages" that are all the same distance long:

If each stage is 11 light-years long, what could you write for how far the spaceship goes in *y* stages?

Fig. 3.1

Wendy: There's a letter there.

Interviewer: What does the letter mean?

W: It's telling you how many stages.

I: Right. Can you write anything for how far the spaceship goes?

W: What, shall I write down what I would do? [*Writes* "If *y* was a number I would times it by 11."]

I: Now can you write that without using words, using mathematics instead?

W: What, how'd you mean, like 11 times *y*?

I: Yes, OK.

W: [*Writes* "11 × *y*."] Is that it?

I: Fine.

W: What, is that all it was? Why didn't you say so? I thought you wanted the answer.

I: Do you mean a particular number?

W: Yes!

I: Well, is there a particular number answer?

W: No! Not unless you know what *y* is.

I: Well, then, how could you give me a particular number answer?

W: Well, you can't, but I didn't know you only had to put *that*!

(Booth 1984, pp. 35–36)

Consequently, students may resort to various strategies (like fourteen-year-old Michelle) in order to derive a numerical answer (see fig. 3.2).

What can you write for the perimeter of this shape:
Part of the shape is not drawn.
There are *n* sides altogether, each of length 2.

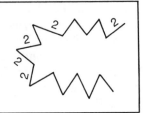

Fig. 3.2

I: Can you write down anything for the perimeter?

Michelle: No, because *n* probably stands for a number and if that's *n*, I can't get how many sides that is, and I've got to add up all the sides, all the 2's, so I need to know how many. Unless *n* stands for . . . like say *n*s in the alphabet, somewhere along the numbers . . . if *n* stands for one of those, then you can say what *n* is.

I: How do you mean?

M: Well, say *n*'s 14.

I: How did you get that?

M: Same as I said before, to get *n* for a number, I got 14. It's 14 along.

I: Oh, did you count along the alphabet?

M: Yes, so that's 14, so I took that for *n*, so it should be . . . 28, to make up the quantity, the perimeter.

I: 28?

M: Yes, you add up all the 2's.

I: Could *n* be another number, or does it have to be 14?

M: Well it could be any *n* number, but if you write the question out like that, without any indication of the number, then there's nothing. . . . The only thing you can do is go along the alphabet and take whatever number *n* comes under to tell you how many 2's you need.

(Booth 1984, p. 33)

Even students who *do* produce a correct algebraic expression (like sixteen-year-old Marie when faced with the problem in fig. 3.3) may not view it as a "proper" answer (an observation also particularly noted by Chalouh and Herscovics [1984]).

What can you write for the perimeter of this shape:
Part of the shape is hidden.
There are *n* sides altogether, each of length 5.

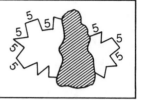

Fig. 3.3

Marie: *n* times 5. It's the number of sides times how long each side is, only you don't know how many sides, so all you can do is *n* times 5.

I: So the answer is *n* times 5?

M: Well, you can't give a proper answer, because you don't know what *n* is. If I knew *n*, I could work it out, but as it is, all you can put is 5*n*.

(Booth 1984, p. 35)

Other students (like fourteen-year-old Michael in fig. 3.4) appear to accept the possibility of an algebraic answer, but they tend to assume that at least what is required is a "single term" answer:

West Ham scored *x* goals and Manchester United scored *y* goals. What can you write for the number of goals scored altogether?

Fig. 3.4

Michael: I can't see any way really, I mean, *x* and *y*, it's really hopeless. Unless, of course, you put above it *x* = 3 or *y* = 2. I've seen that sometimes. And then they put down the question, and you put down the answer.

I: If you did know what the *x* and *y* were, what would you have to do to get the answer?

M: I'd have to add the *y* goals onto the *x* goals, and that would be, say, *z* goals.

I: So you'd say . . . *y* . . .

M: Plus *x*. Yes, so therefore, if they left it like that, I suppose I'd better put down the *z* really. [*Writes just "z."*]

(Booth 1984, p. 20)

Here, Michael's answer nicely demonstrates his appreciation of closure in mathematical systems, even if it is not usually regarded as correct by his teachers! The same idea about single-term answers seems to underlie the commonly observed error by which students "simplify" an expression such as 2*a* + 5*b* to 7*ab*. This problem may occur because students have a cognitive difficulty in "accepting lack of closure" (Collis 1975), or it may simply reflect expectations derived from arithmetic concerning what "well-formed answers" (Matz 1980) are supposed to look like.

There is also another aspect to this problem: Not only are unclosed algebraic expressions legitimate as "answers," but the expression may represent the procedure or relationship by which the answer was obtained as well as the answer itself. For example, "$n + 3$" can be both an "instruction" (or procedure) statement, which states that 3 is to be added to the variable n, and an "answer," which gives the result of having performed the addition. In the first instance, the expression can be interpreted as "add 3 to n"; in the second, as "the number that is 3 bigger than n." This "name-process dilemma" (Davis 1975) can be a source of considerable difficulty for students. However, this problem may relate more closely to the difficulty that students appear to have in accepting algebraic answers than to the fact that the same expression represents both procedure and answer.

NOTATION AND CONVENTION IN ALGEBRA

Students' Interpretation of Symbols

Part of the problem in students' attempts to simplify expressions such as $2a + 5b$ concerns their interpretation of the operation symbol. In arithmetic, symbols such as + and = are typically interpreted in terms of actions to be performed, so that + means to actually perform the operation and = means to write down the answer (Behr, Erlwanger, and Nichols 1980; Ginsburg 1977). Such an interpretation does not seem to be restricted to elementary school children. Thus Kieran (1981) showed within the context of equations that twelve-to-fourteen-year-old students typically regard the equals sign as a unidirectional symbol preceding a numerical answer, as did Wagner's (1977) seventeen-year-old students. The idea that the addition symbol may signal the result of addition as well as the action, or that the equals sign can be viewed as indicating an equivalence relation rather than a "write down the answer" signal, may not be readily appreciated by the student, although both notions are necessary to algebraic understanding. The restricted way of reading the operation symbol also underlies the "name-process dilemma" already described.

In examples such as $2a + 5b$, the actual action associated with the addition symbol is most often to conjoin the terms, producing $7ab$ as an answer. This is perhaps not surprising, given that early ideas of addition involved the physical conjoining of two sets. Furthermore, the fact that the conjoining of terms to denote addition does appear in arithmetic—in mixed fractions (e.g., $2\frac{1}{2} = 2 + \frac{1}{2}$) and also implicitly in place value (e.g., $43 = 4$ tens + 3 units)—may lead students to view the situation similarly in algebra (Matz 1980). Certainly there is evidence to show that children who have never studied algebra before show a strong tendency to "simplify" an expression such as $a + b$ to ab (Booth 1984).

Confusion with the "place value" aspect of conjoining may lead to another kind of error, as evidenced by fifteen-year-old Wayne:

> *Wayne:* [*Explaining the meaning of y in* "add 3 to 5y."] *y* could be a number, it could be a 4, making that (5y) 54 . . . [*Writes* 54].
>
> *W:* [*In response to a request to read out what he has written.*] What . . . five four . . . no, fifty-four!
>
> *I:* Fifty-four?
>
> *W:* Yes.

Choosing a two-digit number to replace the single letter may create some conflict, as in this interpretation:

> *I:* [*Continues the interview above.*] Could *y* be anything else, besides 4?
>
> *W:* Yes, 7, 8, anything!
>
> *I:* So *y* could be any number? [*W nods.*] Suppose I made it 23. What would you write down then?
>
> *W:* Oh! [*Laughs.*] Well! [*Laughs again.*] Five hundred and twenty-three! But I dunno—it doesn't sound very promising! I dunno. Wait, it could be 28, 5 plus 23 . . . yes. . . . [*Writes* 5 + 23.]

(Booth 1984, p. 31)

These findings would seem to lead to several suggestions. The first relates to the ideas concerning the meaning of the operation and equals symbols that children acquire during their early arithmetical experiences. Thus children need to be made aware that "2 + 3" represents not only an *instruction* to add 2 and 3 but also the result that is obtained by performing the addition. This could be done, for example, by reading the expression not only as "2 plus 3" or "add 2 and 3" but also as "the number that is 3 more than 2." Similarly, the bidirectional value of the equals sign needs to be stressed, both by requiring an appropriate reading of the symbol (e.g., "is equal to" rather than "makes," as in "2 plus 3 makes 5") and by giving pupils experience with expressions of the form $5 = 2 + 3$ (as well as $1 + 4 = 2 + 3$, etc.).

The second suggestion relates to the introduction of the conjoined term to represent multiplication in algebra (e.g., $3n$). The apparently strong tendency that children have to view this as the sum rather than the product (or as a place-value representation) would seem to indicate that its introduction should be delayed and that the product should be written in full ($n \times 3$ or $3 \times n$) for a substantial period of the students' early work in algebra. Even when the abbreviated form is introduced, it may be wise to continue to write the product in its expanded form as well, at least over the introduction period.

Third, the "2 apples plus 5 bananas" approach to the problem of $2a + 5b$ also may not be helpful. Not only does it encourage an erroneous view of the meaning of letters, but it can also be used by students to justify their simplification of $7ab$. Thus in the SESM research, the same number of students used this illustration to explain why $2a + 5b$ was equal to $7ab$ (2 apples plus 5 bananas is 7 apples-and-bananas) as those who used it to explain why $2a + 5b$ could not be simplified further.

The Need for Notational Precision

Another area that is more critical in algebra than in arithmetic is the need for exact precision in the recording of statements. Such precision is, of course, also important in arithmetic, but inadequacies in this regard can be of smaller consequence if the student knows what is intended and performs the correct operation regardless of what is written. In arithmetic, whether the student writes $12 \div 3$ or $3 \div 12$ makes little difference if the subsequent computation is correct. However, in algebra the distinction between $p \div q$ and $q \div p$ is crucial. This apparent lack of rigor may reflect a lack of attention in mathematics classrooms to the correct and precise verbal statement of ideas in mathematics. More notably, the free interchange of expressions such as "$12 \div 3$" and "$3 \div 12$" can often be traced to the students' earlier experiences in arithmetic. Some students think that division, like addition, is commutative. Others see no need to distinguish the two forms, believing that the larger number is always divided by the smaller. This appears to derive from well-intentioned advice given by the arithmetic teacher when they first started learning division and from the students' own experience, in which all the division problems met in elementary arithmetic did in fact require the larger number to be divided by the smaller. Presumably this particular misconception could be avoided by the early introduction of problems where the smaller number *is* to be divided by the larger, as when one cake is to be shared equally among six children—a situation with which even very young children are familiar. If not attended to, misconceptions of this kind in arithmetic can lead to problems later in algebra.

LETTERS AND VARIABLES

Letters in Algebra

One of the most obvious differences between arithmetic and algebra, of course, is in the latter's use of letters to represent values. Letters also appear in arithmetic, but in quite a different way. The letters "m" and "c," for instance, may be used in arithmetic to represent "meters" and "cents," rather than representing *the number of* meters or *the number of* cents, as in algebra. Confusion over this change in usage may result in a "lack of numerical referent" problem in students' interpretation of the meaning of letters in algebra, as with fifteen-year-old Peter:

I: And what does the y mean, in a question like that [*Add* 3 *to* 5y.]? Does it mean anything, does it stand for anything, or is it just a letter, or what?

Peter: It's a letter, but it stands for something. It means eight lots of *y*.

I: And what is the *y*?

P: Could be anything.

I: Like what?

P: Could be . . . a yacht. Could be eight yachts.

I: OK, anything else?

P: Could be yoghurt. Or a yam.

I: Would it have to begin with *y*, like yoghurt, or could it be something else?

P: Think it would have to begin with *y*, 'cause you've got a letter *y* there. So you need a *y* for the start of the word.

(Booth 1984, p. 28)

This interpretation may seem a little astonishing, and yet perhaps it should not surprise us. For example, "3 m" is read in arithmetic as "3 meters," and statements such as "3 m = 300 cm" are interpreted as "3 meters are equivalent to 300 centimeters." In algebra, too, reading the variable letters as "labels" in this way often appears to be correct. Reading the statement "$a = l \times w$" as a shorthand version of the verbal statement "area = length $\times$ width," for example, is perhaps hard to distinguish from a reading of the statement as a relationship between the appropriate measures or variables. The apparent correctness of the literal reading of the algebraic statement in this instance may well encourage students to do likewise with terms of the "$5y$" kind.

It is difficult to find ways around this problem. There are clearly good reasons for choosing the particular letters *a, l,* and *w* in representing the formula for the area of a rectangle. However, perhaps this kind of alliteration is best avoided, at least initially. Thus care needs to be taken over the introduction of the kind of example that states "*a* represents the number of apples," leaving the student to translate "$3a$" as "3 apples," rather than as "3 *times* the *number of* apples." (The writing of the product in full as suggested earlier will also help to avoid this kind of mistranslation.) Similarly, in the writing of the rule for the area of a rectangle, students could be left to decide on their own the variables to use in the rule statement. In the SESM research, where students did in fact choose their own letters in such circumstances, it was interesting to note that virtually none of them elected to use the initial letters. Indeed, there was no real reason why they should have done so, since they were choosing variables to represent numerical values that might give the lengths of sides of different rectangles. They were not, therefore, necessarily thinking of the words *length* and *width*. Of course, when students are familiar with this use of letters, the advantage (as a mnemonic device) of using particular letters as variables in a particular situation can be discussed.

The Notion of "Variable"

Perhaps one of the most important aspects of algebra, however, is the idea of "variable" itself. Even when children do interpret letters as representing numbers, there is a strong tendency for the letters to be regarded as standing for specific unique values, as in "$x + 3 = 8$," rather than as numbers in general or variables, as in "$x + y = y + x$" or "$a = l \times w$" (Kuchemann 1981). In arithmetic, symbols representing quantities always do signify unique values. There is little choice, for example, concerning the value represented by the symbol "3." Consequently, it is perhaps not strange that children should treat these new symbols as representing quantities in a similar manner.

One problem arising from this view of letters is that children often assume that different letters must therefore stand for different numerical values. Consequently, many students consider that "$x + y + z$" can never be equal to "$x + p + z$," as with fifteen-year-old Trevor when confronted with the problem in figure 3.5.

$$x + y + z = x + p + z$$

Is this statement true? Always/Never/Sometimes, when. . . .

Fig. 3.5

Trevor: It won't be true, never.

I: Never?

T: Never, because it'll have different values . . . because p has to have a different value from y and the other values, so that'll never be true.

I: So p has to have a different value . . . why do you say that?

T: Well, if it didn't have a different value, then you wouldn't put p, you'd put y. You see, you put a different letter for every different value.

Tristan, also fifteen years old, justifies his opinion that y and p are different values:

Tristan: . . . y couldn't be the same as p.

I: Oh, I see, so using different letters . . .

T: Means they're different amounts.

I: Oh, I see. And are they always different amounts?

T: Well, I've always found they're different. I've never come across one where they're the same.

(Booth 1984, pp. 14–15)

In fact, Tristan almost certainly has come across "one where they're the same," namely, the simple linear graph $y = x$. However, the apparent contradiction provided by this example may well not have bothered him.

The typical response from a class of fifteen-year-olds faced with this problem was that you *could* have $y = x$ here, since "this is graphs, not algebra"!

STUDENTS' UNDERSTANDING OF ARITHMETIC

Many of the difficulties so far described have been discussed from the perspective of the differences between arithmetic and algebra. However, algebra is not separate from arithmetic; indeed, it is in many respects "generalized arithmetic." And herein lies the source of other difficulties. To appreciate the generalization of arithmetical relationships and procedures requires first that those relationships and procedures be apprehended within the arithmetical context. If they are not recognized, or if students have misconceptions concerning them, then this may well affect the students' performance in algebra. In this event, the difficulties that students experience in algebra are not so much difficulties in algebra itself as problems in arithmetic that remain uncorrected.

The Misunderstanding of Arithmetical Conventions

One area where students' ideas on arithmetic can influence their performance in algebra is in the use of parentheses. Children typically do not use parentheses (Kieran 1979), because they believe that the written sequence of operations determines the order in which the computation should be performed. In addition, many students think the value of an expression remains unchanged even if the order of calculation is varied:

> [*Keith, thirteen years old, computing* $18 \times 27 + 19$, *having just calculated* $27 + 19 \times 18$ *from left to right*]
>
> *Keith:* Do . . . 27 plus 19, then multiply by 18. It's the same as the last one . . . it's just the other way around.
>
> *I:* Right, well, suppose I came along and thought it meant multiply 18 by 27, and then add 19. Would I get the same answer?
>
> *K:* Yes.
>
> *I:* Which way would you do it?
>
> *K:* Either! Either way. Depends what comes into my mind at the time.
>
> *I:* But would it matter which way you did it?
>
> *K:* No, you'd still get the same answer.
>
> (Booth 1984, p. 55)

A further view is that the context to which the written expression relates will determine the order of computation regardless of how the expression is written, as fifteen-year-old Neil thought (fig. 3.6):

What can you write for the area of this rectangle: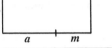

Fig. 3.6

Neil: p times . . . *a* plus *m*. [*Writes p × a + m.*]
I: Right, so you've written down *p × a + m*. And what would you actually do, what would you need to do first?
N: I'm not with you.
I: Right, why did you say *p* times *a* plus *m*?
N: Because you're timesing that side [*a and m*], and that side [*a and m*] you can't do, so you've got to add that [*a*] onto that [*m*], to times the two sides together.
I: Right, so which bit would you do first?
N: . . . I'd add those two up [*a and m*], and then I'd times it by *p*.
I: And is that what you've written?
N: Yes.
I: Suppose I said I thought that [*p × a + m*] meant *p* times *a*. And then plus *m*.
N: Oh no, it can't be that. If you did *p* times *a,* you'd only get a bit of it [*area*]. You've got to do the *a* plus *m* to get the whole length, and then times *p*. You've got to add *a* and *m* first.

(Booth 1984, p. 22)

In these examples, the results are to ignore the need for parentheses. Consequently, algebraic expressions requiring parentheses are incorrectly written (e.g, $p \times a + m$ instead of $p \times (a + m)$), which may result in further errors when the expression is simplified (e.g., $p \times a + m$ may then be rewritten, erroneously in this context, as $pa + m$). Here the error arises not so much from an algebraic misconception as from an incorrect view of arithmetical representation.

Students' Use of Informal Methods

There is considerable evidence that children at the elementary school level use informal problem-solving methods (e.g., Ginsburg 1975; Carpenter and Moser 1981), and a similar observation has also been made at the secondary school level (e.g., Booth 1981; Petitto 1979). Where the solution of equations is concerned, for example, the availability of only informal procedures can have a marked effect on students' facility with seemingly similar items. In a study involving approximately two hundred sixteen-year-old students representing the top 20 percent of the ability range in a sample of Swedish school students, Ekenstam and Nilsson (1979) found that although 82 percent of the students solved the equation $30/x = 6$ correctly, only 48 percent were successful with the structurally similar example $4/x = 3$. In the first example, students were able to solve the equation by inspection, a procedure that could not be so readily applied in the second example.

The use of informal methods in arithmetic can also have implications for students' ability to produce (or understand) general statements in algebra. For example, if a student typically does not find the total number of elements in two sets of, say, 35 and 19 elements by using the notion of addition as represented by $35 + 19$ but rather solves the problem by a counting-on procedure, then the chances are perhaps slight that the total number of elements in two sets of x and y elements will be readily represented by $x + y$. Here the difficulty is not so much one of generalizing from the arithmetic example as it is of having an appropriate procedure, and a representation of that procedure, in arithmetic from which to generalize in the first place. This whole question of the importance of the use by students of their own informal methods has been discussed elsewhere (e.g., Case 1974; Booth 1981; Booth and Hart 1983). If students are to learn (and use) the more formal procedures, they must first see the need for them. This requires (a) that the teacher recognize that students may have an informal method for a given kind of problem; (b) that the value of this informal method for the solution of simple problems be recognized and discussed; and (c) that the possible limitations of the method be considered, by the simple process of attempting to use it to solve a harder problem of the same kind. By this means it is suggested that students' recognition of the need for a more general (i.e., formal) procedure can be obtained. Ways of helping the students develop an understanding of the formal procedure itself must then be found.

IN CONCLUSION

This list of possible causes of children's difficulties in learning algebra is by no means exhaustive. It may, however, serve to throw some light on the kinds of difficulty that children are likely to experience when they begin to study algebra. Since the value of such insights must lie in the use to which they can be put in making decisions concerning the teaching and learning of algebra, we must ask what the teacher can do in order to help children avoid or correct these problems. It is hoped that the suggestions made here may go some way toward meeting this requirement. In addition, the illustrations provided here may serve to remind us that some seemingly simple ideas are not always as simple for students as they may seem to adults. A continuing assessment of exactly what is involved in the learning of new mathematical topics, assisted by an analysis of the errors that students make and the reasons for them, may provide us with extremely useful tools for deciding on ways to help children improve their understanding in mathematics. It remains for teachers and researchers to take what steps they can to further this endeavor.

REFERENCES

Behr, Merlyn, Stanley Erlwanger, and Eugene Nichols. "How Children View the Equals Sign." *Mathematics Teaching* 92 (1980): 13–15.

Booth, Lesley R. *Algebra: Children's Strategies and Errors.* Windsor, England: NFER-Nelson, 1984.

————. "Child Methods in Secondary Mathematics." *Educational Studies in Mathematics* 12 (1981): 29–40.

Booth, Lesley R., and Kathleen Hart. "Doing It Their Way: Some Child-Methods in Mathematics." In *Research Monograph of the Research Council for Diagnostic and Prescriptive Mathematics*, pp. 80–84. Kent, Ohio: RCDPM, 1983.

Carpenter, Thomas P., and James M. Moser. "The Development of Addition and Subtraction Problem-solving Skills." In *Addition and Subtraction: Developmental Perspective*, edited by Thomas P. Carpenter, James M. Moser, and Thomas Romberg. Hillsdale, N.J.: Lawrence Erlbaum Associates, 1981.

Case, Robbie. "Mental Strategies, Mental Capacity, and Instruction: A Neo-Piagetian Investigation." *Journal of Experimental Child Psychology* 82 (1974): 372–97.

Chalouh, Louise, and Nicolas Herscovics. "From Letter Representing a Hidden Quantity to Letter Representing an Unknown Quantity." In *Proceedings of the Sixth Annual Meeting of the Psychology of Mathematics Education, North American Group*, edited by James Moser. Madison, Wis.: University of Wisconsin, 1984.

Collis, Kevin F. *A Study of Concrete and Formal Operations in School Mathematics: A Piagetian Viewpoint.* Melbourne: Australian Council for Educational Research, 1975.

Davis, Robert B. "Cognitive Processes Involved in Solving Simple Algebraic Equations." *Journal of Children's Mathematical Behaviour* 1(3) (1975): 7–35.

Ekenstam, Adolf A., and Margita Nilsson. "A New Approach to the Assessment of Children's Mathematical Competence." *Educational Studies in Mathematics* 10 (1979): 41–66.

Ginsburg, Herbert. *Children's Arithmetic: The Learning Process.* New York: Van Nostrand, 1977.

————. "Young Children's Informal Knowledge of Mathematics." *Journal of Children's Mathematical Behaviour* 1(3) (1975): 63–156.

Kieran, Carolyn. "Children's Operational Thinking within the Context of Bracketing and the Order of Operations." In *Proceedings of the Third International Conference for the Psychology of Mathematics Education*, edited by D. Tall, pp. 128–32. Warwick, England: University of Warwick, 1979.

————. "Concepts Associated with the Equality Symbol." *Educational Studies in Mathematics* 12 (1981): 317–26.

Kuchemann, Dietmar E. "Algebra." In *Children's Understanding of Mathematics: 11–16*, edited by K. Hart, pp. 102–19. London: Murray, 1981.

Matz, Marilyn. "Towards a Computational Theory of Algebraic Competence." *Journal of Children's Mathematical Behaviour* 3(1) (1980): 93–166.

Petitto, Andrea. "The Role of Formal and Informal Thinking in Doing Algebra." *Journal of Children's Mathematical Behaviour* 2(2) (1979): 69–88.

University of Bath. *Mathematics in Employment: 16–18.* Bath, England: University of Bath, School of Mathematics, 1982.

Wagner, Sigrid. "Conservation of Equation, Conservation of Function, and Their Relationship to Formal Operational Thinking." Doctoral diss., New York University, 1977.

4

Teaching Algebraic Expressions in a Meaningful Way

Louise Chalouh
Nicolas Herscovics

QUITE OFTEN, algebraic expressions are introduced by stating that they involve variables and that "a variable is a letter that stands for one or more numbers." Such formal definitions may be adequate for mathematics teachers but they often fail to provide meaning for the beginning student. The construction of meaning for algebraic expressions by novices necessitates finding in their background a cognitive basis on which to build. Helping students create meaning on the basis of their existing knowledge has been our prime objective. To achieve it, we designed an innovative teaching outline.

The teaching outline introduces algebraic expressions as *answers to problems*. The problems chosen involve previously learned concepts such as the number of dots in a rectangular array, the length of a line segment, and the area of a rectangle. The advantage of these types of problems is that they have an easy visual representation. The use of letters is introduced cautiously by first letting them represent *hidden* quantities and only afterwards using them to stand for specific *unknown* quantities. This geometric approach is used systematically to help students construct meaning for expressions involving one unknown and one operation to expressions with several unknowns and multiple operations. Not only are the students learning to use algebraic expressions as answers to problems, but they are also encouraged to reverse the process—that is, they are asked to generate problems corresponding to given expressions.

Our teaching outline, which covers three lessons, has been tested by teaching six students individually. This teaching experiment was intended to determine the accessibility of our new presentation as well as to uncover any new cognitive problems we may have inadvertently introduced. Individual

33

instruction enabled us to get immediate feedback from the students by gathering their spontaneous responses and by further probing their thinking, thus providing us with information often unavailable in a classroom.

In designing our teaching outline, we took into account four cognitive obstacles in the learning of algebraic expressions that had been identified in prior research. The first was the student's *lack of a numerical referent* for the letter used (Davis 1975; Wagner 1981). If the learner does not view letters as representing numbers, then performing arithmetic operations with them is a meaningless task. Another obstacle more specific to algebraic expressions is their perception as incomplete statements. Collis (1974) explained this in terms of the student's inability to hold unevaluated operations in suspension. In arithmetic "2 + 3" can be replaced by "5," but an expression such as "$x + 3$" cannot be replaced by another number. Collis referred to the student's difficulty in holding unevaluated operations in suspension as their *inability to accept the lack of closure.* Davis (1975) pointed out another cognitive obstacle, the *name-process dilemma,* that distinguishes algebra from arithmetic. In the latter, "2 + 3" is the problem and "5" is the answer, whereas "$x + 3$" both describes the *process* (adding 3 to x) and *names* the answer. Matz (1979) has described another obstacle, that of the different meaning associated with *concatenation* in algebra. In arithmetic, the juxtaposition of two numbers denotes addition ($43 = 40 + 3; 4\frac{1}{2} = 4 + \frac{1}{2}$); in algebra, concatenation denotes multiplication ($4a = 4 \times a$).

THE TEACHING OUTLINE

For our teaching outline, three types of problems were selected on the basis of their easy geometric representation (see fig. 4.1). The first type of problem dealt with determining the total number of dots in a rectangular array for which only one dimension (a row or column) was shown; the second type was finding the length of a line segment in which the number of parts was hidden; the third type involved finding the area of a rectangle with only one dimension shown. In all cases, the hiding was done by using a cardboard cover.

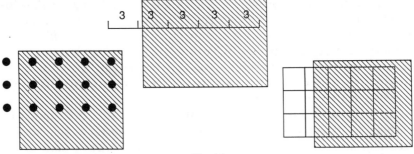

Fig. 4.1

In our teaching experiment our pupils were three sixth-grade students (Frankie, Wendy, and Antoinetta) and three seventh-grade students (Yvette, Filippo, and Gail), none of whom had had any prior introduction to a formal course in algebra. They were selected by the school authorities on the basis of their scholastic performance in order to provide us with a weak, average, and strong pupil from each of these grades. The students were pretested to make sure that they could handle the three types of problems at the numerical level and that they perceived their multiplicative nature. Each student was taught individually to provide us with immediate feedback concerning any difficulty he or she was experiencing. Because of spatial constraints, we shall present mainly the area problems, except for Lesson 1, where the introduction was based on the dot problems.

Lesson 1: From Placeholder to Letter Representation

In order to make the introduction to algebraic expressions as simple as possible, pupils were initially taught to use the familiar box as a placeholder for the partially hidden dimension. At first they were presented with a 5 × 7 array of dots in which only the first row was exposed (fig. 4.2):

In this exercise you can see a row of 7 dots. And I have hidden more rows, each with 7 dots. Here is the problem. How would I write the total number of dots if I don't know the exact number of rows?

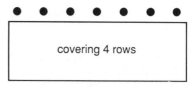

covering 4 rows

So let me show you how I do it:

Number of dots = 7 × □

Fig. 4.2

Since I don't know the exact number of rows, I am meanwhile using a box. Now I am going to let you have a peek at what I have covered and ask you to fill in the right number in the box.

The use of the box did not appear troublesome for any of the students, except initially for Wendy, our average sixth-grader, who felt the expression was incomplete and had to provide a numerical value for the box when she saw it. All six pupils wrote the number 5 inside the box, thus indicating that they correctly perceived the box as representing the total number of rows, not just those that were covered. Given another such dot problem, all students were able to complete the statement "Number of dots = ____" by writing "8 × □." Finally, they were given an expression and asked to generate a dot problem:

Now it is your turn. Can you make up a dot problem like we just did, where 5 × □ is the total number of dots?

A typical response was that of Gail, our strong seventh grader, who hid her work from the interviewer until she turned around, showing a row of five dots while covering the remaining rows. The other five pupils behaved in a similar manner, thus indicating that they had grasped the intended meaning of the box as representing an undetermined quantity that was partially hidden.

Following this initial work with the box as a placeholder, the problems were reintroduced with the only change being a request that the student use a letter of his or her choice (instead of a box) to represent the hidden quantity. This request originated from the need to convey the arbitrary nature of the symbol. The first problem was presented as follows (see fig. 4.3):

Let's do another dot problem. You can see that I have a row of 5 dots, but you can't see the exact number of rows. This time I am going to use a letter instead of a box for the number of rows. Choose a letter. So let me show you how I write the total number of dots using your letter: Number of dots = a [or whatever letter is chosen].

Now I will let you look at the number of rows. How many rows are there? What does your letter stand for? Can you complete this equation?

Number of dots = 5 × ____

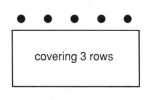

Fig. 4.3

All the students appeared to accept the use of the letter they chose as part of the algebraic expression representing the total number of dots. As before with the boxes, they exhibited no difficulty in understanding that the letter stood for the total number of rows. The presentation used here provided a situation where the letter became a natural extension of the students' use of the box as a placeholder, and both were explicitly linked to a numerical referent. When questioned, all six pupils stated that it did not bother them to see a number multiplying a letter, thus confirming that no cognitive dilemma existed for them and suggesting that the primary aim of Lesson 1 had been achieved.

Lesson 2: From Letter Representing a Hidden Quantity to Letter Representing an Unknown Quantity

Since our pupils could now use literal symbols without experiencing any difficulties in the context of a *hidden* quantity, they were ready to be introduced to the use of the literal symbol as representing an *unknown* quantity. Since the problem types were the same as in the previous lesson, it was presumed that the use of a letter representing an unknown would be just an easy extension of its previous use in representing a hidden quantity. Thus we believed the pupils would have no difficulty in answering the first question in the second lesson: "Do you think you could write down the area of this rectangle?" (See fig. 4.4.) All that was required was for the student to write

"8 x *a*." It should be noted that the form of this question differed from that of the problems in Lesson 1, where each question was started by requesting the student to complete a statement such as "Area of rectangle = ____."

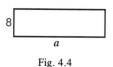

Fig. 4.4

The responses given by our pupils to this first problem highlighted some unforeseen difficulties. All of them misinterpreted the question in the sense that they thought they had to provide a *numerical* answer to the problem. This need to provide a numerical answer did not manifest itself in Lesson 1. An explanation for the difficulty in this transition could be the students' inability to accept the lack of closure. Two of our pupils, Wendy and Gail, indicated that they viewed the expression "8 × *a*" to be incomplete by writing an equal sign after the expression. Both Frankie and Yvette wrote the expression "8 × *a*" but immediately made some reference to evaluating *a*. Antoinetta and Filippo were not able to write the expression, whereas they had managed to do so in the previous lesson. All these responses were indications that our students could not accept the algebraic expression by itself.

However, they showed that it was far more acceptable to them in the context of completing the statement "Area = ____." A possible explanation of the students' greater acceptance of the expression within the framework of an equation can be sought in terms of the *name-process dilemma*. When they complete the equation "Area = ____," the left-hand side clearly expresses to them the *name*, thus leaving to the right-hand side the function of expressing the *process*, "8 × *a*." Filippo expressed this idea most explicitly by stating, "You just have to say what you have to multiply to get the area."

A second problem provided another situation for our pupils to write an algebraic expression to represent the area of a rectangle. All of them provided an algebraic answer: "Area = 3 × *c*" (two students just wrote "3 × *c*"). For the third question, students were asked to generate an area problem that would correspond to the expression "5 × *d*." All pupils succeeded except Wendy, who initially drew the diagram shown in figure 4.5. When it was pointed out to her that *d* was not an unknown quantity in her diagram, she was able to correct herself.

Fig. 4.5

In order to maintain the numerical referent for the literal symbol that had been established in Lesson 1, the students were asked to measure the length of the unknown dimension in the first area problem. The measurement was followed by the substitution of this numerical value in the algebraic expression representing its area.

THE INTRODUCTION OF CONCATENATION

Students' problems in using concatenation are easily underestimated. This is why we dealt with it quite carefully. The introduction began as follows:

> How do you write 6 times 7 in arithmetic? In algebra we can use all the letters in the alphabet, capitals and small letters. We often use the letter x. This can get a little confusing, especially if you have two x's, as in "$8 \times x$" (8 times x). In order not to have any confusion, in algebra we can just leave out the multiplication sign and write "$8x$" for "$8 \times x$."
>
> Whenever you see a letter attached to a number, there is a hidden multiplication there. What does $5a$ mean? What does $4b$ mean?

The presentation then proceeded to illustrate why this type of notation was only possible in algebra:

> Can we do this in arithmetic? Can we remove the multiplication sign in 6×7 and write the numbers next to each other?

Students were then asked to make up an area problem corresponding to $5a$. They had no difficulty with the concatenated form and were introduced to two more conventions. The first was that of multiplying a letter by the number 1, that is, writing $1x$ as x.

> Here is a rectangle (fig. 4.6). Can you tell me what the area is? (Expected response: $1x$) What do we do with a product like "1×3"? (Expected response: "just write 3") The same thing happens in algebra—we omit the 1 and just write x.

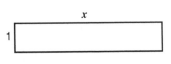

Fig. 4.6

The second convention to be presented specified that in algebra, the product of a number and a letter is always written with the number preceding the letter:

> One other convention that I want to show you is the following one: We write $a \times 3$ as $3a$. Does it make sense? It does not make much difference, since $a \times 3 = 3 \times a$. Is this true for any value of a? Can you check it with some numbers? From now on, we write $b \times 5$ as ____ and $c \times 6$ as ____.

On the basis of their answers, it seemed that all our students accepted these notational conventions without any difficulty.

Lesson 3: Algebraic Expressions with Multiple Terms

Lesson 2 introduced the letter within the context of problems purely multiplicative in nature, but Lesson 3 presented more complex ones involving both multiplication and addition. The first problem (see fig. 4.7) was introduced as follows:

> Here is a rectangle. The length of the height is 3 units. The length of the base is in two parts; one part is unknown, so I marked it c, and the other part is 2 units.

In order to find the area of this rectangle, I'm going to simplify the problem. Here I'm drawing a broken line splitting it up into two smaller rectangles. What is the area of the rectangle on the left? ($3c$) (Write $3c$ in that rectangle.) What is the area of the rectangle on the right? (6) Write 6 in that rectangle.) What is the area of the rectangle we started with?

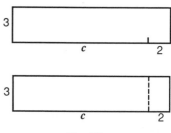

Fig. 4.7

The responses of the students to this problem proved to be very interesting. When asked to write the area of the rectangle on the right, Wendy said "3 times 2" but wrote "32," explaining "three two in algebra." Yvette wrote both "3 × 2" and "32" for "3 times 2." Wendy and Yvette's responses revealed the possibility of students' transferring the concatenation convention from algebra to arithmetic. Although concatenation had been previously presented to the students, three of them did not spontaneously use this convention and had to be reminded about this new form. Another interesting result was that five of the six students initially did not perceive the additive aspect of the problem. After determining the area of the two smaller rectangles to be $3c$ and 6, they responded that the area of the original rectangle was "$3c$ × 6" or "$3c$ × 2." These answers suggest that the area concept is so strongly linked with multiplication that it creates a "mind-set" that prevents the student from perceiving the additive nature of the problem involving the sum of two areas. This strong association between area and multiplication was further evidenced when a purely numerical area problem was presented to the students.

Given the rectangle in figure 4.8 in which the areas of the subrectangles were inscribed, three pupils still used multiplication, 40 × 12, to express the area of the outer rectangle! This problem was finally resolved when the students had to find the area of a rectangle divided into unit squares (fig. 4.9).

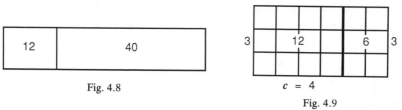

Fig. 4.8

Fig. 4.9

They were able to handle the area problem in figure 4.10 with ease. However, when they were asked to generate a rectangle whose area would be $3a + 9$, five of them drew the rectangle shown in figure 4.11. Although this shows that they had overcome the strong link between area and multipli-

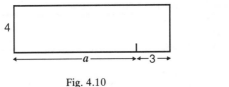

Fig. 4.10

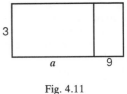

Fig. 4.11

cation, it also indicates that they were focusing on the second term of the expression as being part of the base. Each time they were asked to determine the area of each subrectangle, they were able to correct their error.

The first area problem introducing two unknowns had the base of the rectangle divided into two parts (fig. 4.12):

Here is a rectangle. Can you write the area of this rectangle? (If necessary, have the students determine the area of the subrectangles.)

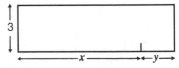

Fig. 4.12

The initial responses indicated that the students were confusing the additive and multiplicative aspects of this problem. Wendy, Frankie, and Antoinetta first wrote "Area $= 3x + y$"; Gail wrote "$3xy$"; Yvette wrote "$3 + xy$." Even after partitioning the original rectangle and inscribing $3x$ and $3y$ in the respective subrectangles, four of these five students multiplied $3x$ by $3y$. Two of them corrected themselves immediately without any help, but the other two had to be reminded of the area problems they had solved previously.

The second area problem (see fig. 4.13) went one step further. The base consisted of three parts, two segments of unknown length and one of known length. All six pupils answered with ease, indicating that they had overcome their initial difficulties. They were then asked to make up an area problem where the answer would be $5d + 5e + 10$. Four of the students drew the diagram shown in figure 4.14. This shows, as in a previous, similar problem, that they perceived the term 10 as part of the length of the base, rather than the area of one of the subrectangles. As before, this was easily corrected.

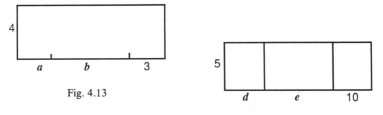

Fig. 4.13

Fig. 4.14

CONCLUDING REMARKS

The instruction we actually provided was more extensive than the short description given here. In each lesson, students were given many more problems involving dot arrays, line segments, and rectangles. With each type of problem, not only did they have to find the corresponding algebraic expression, but they also had to reverse the process and generate all kinds of problems corresponding to given algebraic expressions. In the first two lessons, they were given homework that was corrected with them at the beginning of the next interview.

The dot array problems were good for Lesson 1 but proved to be cumbersome later on. We were unable to find a way acceptable to the students of representing the idea of an unknown quantity of dots. The line segment type of problem had to be readjusted in Lesson 2 where the unknown quantity was no longer associated with the number of segments but with the unknown length of each part: $\underset{\llcorner a \lrcorner a \lrcorner a \lrcorner a \lrcorner}{}$. This inevitably created some confusion for a couple of students but was easily clarified.

The Impact of Our Teaching Outline

Prior to our teaching experiment, we were interested in finding out how students who had not been exposed to any formal instruction in algebra perceived algebraic symbolism (Chalouh and Herscovics 1983). Thus before starting the three lessons, we asked our six pupils what $3a$ meant to them. The expression was interpreted variously as a subdivision label ("third problem, first part"), in terms of place value (e.g., $3a$ is 30), as a first-letter abbreviation (e.g., $3a$ is 3 apples), and in terms of place-value and alphabetical rank ("$3a$ is 31 because a is the first letter of the alphabet"). Asked what they would get by replacing the letter a in $3a$ by the number 2, five of the six students responded with "32." When requested to "add 4 to $3n$," one student responded with "$4 + 3 = n$," whereas the other five pupils showed a place-value interpretation of the question ($4 + 32$, $4 + 38$). These results show the very natural tendency of the students to interpret an algebraic expression new to them in terms of the only numerical frame of reference they possess at this point, that of arithmetic.

We tested our students one month after the third lesson was completed in order to assess the extent to which their perception of algebraic symbolism had changed. We were most surprised to find that five of the six pupils gave answers similar to the ones they provided in the pretest. However, a chance remark by Wendy alerted us to the fact that another frame of reference was present in her mind. To one of the questions in the posttest she inquired if we wanted her to answer "in algebra?" This led us to ask each student, "In algebra, what does $5b$ mean?" Specifically requested to respond "in algebra," Wendy, Gail, Filippo, and Yvette changed their initial answer to an algebraic one and said "5 times b." In substituting 2 for b, they did not say "fifty-two" as before but now read it as "5 times 2." This change in response is interesting, since it shows that all these students knew the meaning of $5b$ within the context of algebra, but unless specifically requested to do so, they remained in an arithmetic context (Herscovics and Chalouh 1985).

The explicit instruction to answer *in algebra* led to a remarkable shift. This is further evidenced in the answers to the request to "add 4 to $3n$." Whereas their initial responses in the posttest were essentially arithmetic, when jolted into an algebraic frame of reference, five of them answered "$4 + 3n$." These results imply that in early algebra, teachers cannot take a student's answers at face value. They must first reassure themselves that the pupil is well aware of the frame of reference in which he or she is to respond.

By way of conclusion, we wish to assess the value of our teaching outline. The question of its accessibility to most, if not all, students must be a determining factor. Our teaching experiment shows that our six pupils, both strong and weak, whether in grade 6 or in grade 7, all succeeded in constructing meaning for algebraic expressions. In fact, the posttest results show that each of them was able to handle even the most complex problems—those involving two variables and three terms. Granted, they needed some assistance, but then our teaching experiment should not be compared to a classroom situation. A teaching experiment such as this one has as its main objective the identification of cognitive obstacles inherent in the teaching outline. However, a classroom adaptation of this presentation would allow for a slower pace and much more practice.

REFERENCES

Chalouh, Louise, and Nicolas Herscovics. "The Problem of Concatenation." In *Proceedings of the Fifth Annual Meeting of the North American Chapter of the International Group for the Psychology of Mathematics Education,* edited by J. C. Bergeron and Nicolas Herscovics, pp. 153–60. Montreal: Université de Montréal, Faculté de Sciences de l'Education, 1983.

Collis, Kevin F. "Cognitive Development and Mathematics Learning." Unpublished paper prepared for the Psychology of Mathematics Workshop, Centre for Science Education, Chelsea College, London, 1974.

Davis, Robert B. "Cognitive Processes Involved in Solving Simple Algebraic Equations." *Journal of Children's Mathematical Behavior* 1 (1975): 7–35.

Herscovics, Nicolas, and Louise Chalouh. "Conflicting Frames of Reference in the Learning of Algebra." In *Proceedings of the Seventh Annual Meeting of the North American Chapter of the International Group for the Psychology of Mathematics Education,* edited by Suzanne Damarin and Marilyn Shelton, pp. 123–31. Columbus, Ohio: PME-NA, 1985.

Matz, Marilyn. "Towards a Process Model for High School Algebra Errors." Unpublished manuscript, Working Paper 181, Massachusetts Institute of Technology Artificial Intelligence Laboratory, 1979.

Wagner, Sigrid. "Conservation of Equation and Function under Transformations of Variable." *Journal for Research in Mathematics Education* 12 (1981): 107–18.

5

Difficulties Students Have with the Function Concept

Zvia Markovits
Bat Sheva Eylon
Maxim Bruckheimer

I NCLUDING the function concept in precalculus mathematics courses has been advocated from the beginning of this century and maybe even earlier. However, it appeared as an explicit topic in most curricula with the coming of the "new math." After some twenty to twenty-five years, it is appropriate to investigate how the concept has been learned by students.

This article describes the findings of a study in which we investigated how a group of ninth- and tenth-grade students who had studied linear functions and functions in general understood the concept of function. We also investigated their difficulties and misconceptions and tried to determine possible causes and alternative instructional procedures that might help to correct them.

We undertook this investigation in the context to the specific curriculum taught in local schools. However, we concentrated on general aspects that have relevance to any curriculum in which students begin functions. This can be seen from the section immediately following, in which we list the components of the function concept included in the study. The next section describes the difficulties and misconceptions we found and suggests some modifications of the curriculum that may well help to overcome them. Readers can thus compare the findings and suggestions with their own curricula.

COMPONENTS IN UNDERSTANDING THE FUNCTION CONCEPT

The components that form the "understanding of the function concept" in our study are listed in figure 5.1. Each component is accompanied by sample

43

problems used in the investigation. (Note that there may be more than one correct response to an item, and that answers are given in the Appendix.) These components, which are only a part of what might be considered aspects of a general understanding of functions, were established after we analyzed the stages that students pass through (or should pass through) in their first course on functions. We took into account two basic facts:

1. A function is defined by two sets, the domain and the range, and by a rule of correspondence that assigns to every element of the domain exactly one element of the range.
2. Most functions (in the precalculus curriculum) have several representations—graph, algebra, table, and arrow diagram, to name four.

Each component in the figure has two stages—the passive stage (classifying and identifying) and the active stage (doing something or giving examples).

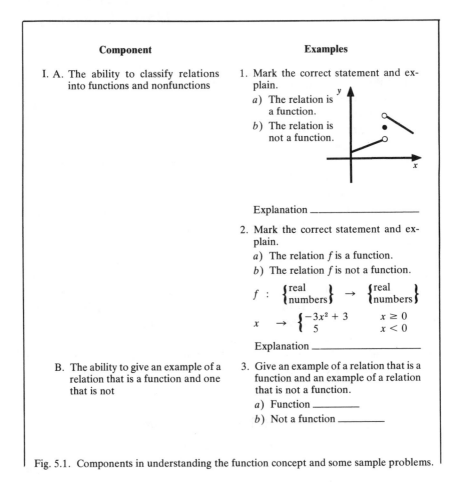

Fig. 5.1. Components in understanding the function concept and some sample problems.

II. A. For a given function, the ability to identify preimages, images, and (preimage, image) pairs

4. For each of the given points and the function represented by the graph, decide if it represents an image, preimage, a (preimage, image) pair, or a point that does not represent a (preimage, image) pair.

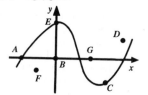

5. Given the function f :

$$f : \left\{ \begin{array}{l} \text{natural} \\ \text{numbers} \end{array} \right\} \rightarrow \left\{ \begin{array}{l} \text{natural} \\ \text{numbers} \end{array} \right\}$$

$$f(x) = 4x + 6$$

a) Which of the numbers
$$2, -1, 0, 11.5, 1267$$
is a preimage of f?

b) Which of the numbers
$$-2, 10, 8, 46, 23$$
is an image under f?

c) Which of the following ordered pairs
$$(5, 26), (0.5, 8), (2, 10)$$
is a (preimage, image) pair of f?

Explain your answers.

B. For a given function, the ability to find the image for a given preimage and vice versa

6. For each of the graphs of the functions given, mark the elements of the range that are images of the point A in the domain.

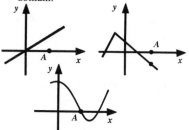

7. For the function g

$$g : \left\{ \begin{array}{l} \text{real} \\ \text{numbers} \end{array} \right\} \rightarrow \left\{ \begin{array}{l} \text{real} \\ \text{numbers} \end{array} \right\}$$

$$g(x) = -7$$

a) Complete the following:

$g(4) =$ ☐ $g(-7) =$ ☐
$g(0) =$ ☐ $g(3.5) =$ ☐

Fig. 5.1 *(continued)*

b) Is there a real number x, such that $g(x) = 3$? How many such x's exist? Explain.

c) Is there a real number x such that $g(x) = -7$? How many such x's exist? Explain.

III. A. The ability to identify equal functions

(Two functions f, g are defined to be equal if they have the same domain and range and if for every x in the domain, $f(x) = g(x)$.)

8. Given the function f

$$f : \begin{Bmatrix}\text{natural}\\\text{numbers}\end{Bmatrix} \rightarrow \begin{Bmatrix}\text{natural}\\\text{numbers}\end{Bmatrix}$$

$$f(x) = 4x + 6$$

For each of the following, decide whether it describes a function equal to f and explain:

a) $g : \begin{Bmatrix}\text{real}\\\text{numbers}\end{Bmatrix} \rightarrow \begin{Bmatrix}\text{real}\\\text{numbers}\end{Bmatrix}$

$$g(x) = 4x + 6$$

b) $g : \begin{Bmatrix}\text{natural}\\\text{numbers}\end{Bmatrix} \rightarrow \begin{Bmatrix}\text{natural}\\\text{numbers}\end{Bmatrix}$

$$g(x) = 2x + 3$$

c)

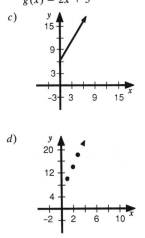

d)

B. The ability to transfer from one representation to another

9. Find the algebraic form of the function shown in the graph, specifying its domain and range.

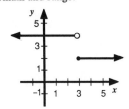

Fig. 5.1 *(continued)*

10. Draw the graph of the function:

$$g : \left\{\begin{array}{l}\text{real}\\\text{numbers}\end{array}\right\} \rightarrow \left\{\begin{array}{l}\text{real}\\\text{numbers}\end{array}\right\}$$

$$g(x) = x - 2$$

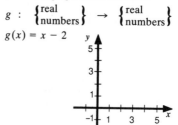

IV. A. The ability to identify functions satisfying given constraints

11. Indicate those graphs that represent a function with domain $\{x \mid 2 < x < 6\}$ and range $\{y \mid -1 < y < 4\}$.

(a)

(b)

(c)

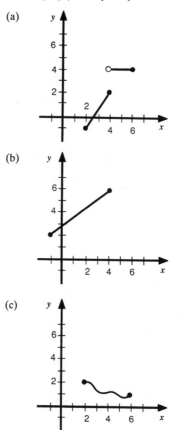

B. The ability to give examples of functions satisfying given constraints

12. a) Give an example in algebraic form of a function from the real numbers to the natural numbers.

Fig. 5.1 (continued)

b) The number of different such functions is—

- 0
- 1
- 2
- more than 2 but fewer than 10
- more than 10 but not infinite
- infinite

Explain your answer.

13. *a*) In the given coordinate system draw the graph of a function such that the coordinates of each of the points *A, B* represent a preimage and the corresponding image of the function.

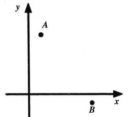

b) The number of different such functions that can be drawn is—

- 0
- 1
- 2
- more than 2 but fewer than 10
- more than 10 but not infinite
- infinite

Explain your answer.

14. *a*) In the given coordinate system draw the graph of a function such that the coordinates of each of the points *A, B, C, D, E, F* represent a preimage and the corresponding image of the function.

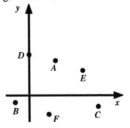

b) The number of different such functions that can be drawn is—

- 0
- 1
- 2

Fig. 5.1 *(continued)*

> - more than 2 but fewer than 10
> - more than 10 but not infinite
> - infinite
>
> Explain your answer.
>
> 15. *a*) Give an example in algebraic form of a function *f* for which
>
> $f(3) = 4$, $f(6) = 7$, and $f(8) = 13$
>
> *b*) The number of different such examples is—
> - 0
> - 1
> - 2
> - more than 2 but fewer than 10
> - more than 10 but not infinite
> - infinite
>
> Explain your answer.

Fig. 5.1 *(continued)*

THE FUNCTION CONCEPT: DIFFICULTIES, MISCONCEPTIONS, AND REMEDIES

The items in figure 5.1 are only examples from a large variety of items dealing with functions in graphical and algebraic form that were used in the study. In this section, we use the examples in the table to describe the difficulties and misconceptions we found and to suggest some possible causes and remedies that we are currently implementing in the revision of the curriculum materials.

Terms

Students often have difficulty with the terms *preimage, image, (preimage, image) pair, domain, range,* and *image set.* This leads to other difficulties, such as locating preimages and images on the axes in the graphical representation, identifying images and (preimage, image) pairs for functions given in algebraic form, distinguishing between the image set and the range, and ignoring the domain and range of a function. Let us discuss each of these further.

Difficulties in locating preimages and images on the axes in the graphical representation. David (grade 10) gave the answer shown in figure 5.2 to problem 6. This was typical of many students who located the images and preimages incorrectly on the graph of the function. They did not appreciate that in the graphical representation the *x*-axis represents the domain and the *y*-axis the range, whereas the points on the graph represent (preimage, image) pairs.

There is here a clear and fundamental difficulty. Many students do not make the connection between the components in the verbal definition of a function and the corresponding components in the visual graphical repre-

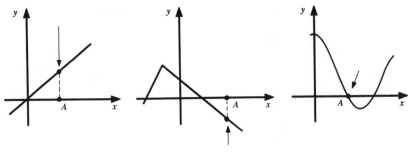

Fig. 5.2

sentation. There is a subsidiary difficulty intrinsic to the graphical form. This is the double role of the points on the axes: They are points in the plane with coordinates $(x, 0)$ or $(0, y)$ and in this form can represent (preimage, image) pairs when the function intersects one of the axes. But they are also points on a number line, and in this form they can represent preimages or images. In all other representations (algebra, arrow diagrams, etc.) we do not have this problem.

In the revision of the curriculum we have tried to overcome these difficulties by direct treatment—that is, we ask questions dealing with the connection between the function components and their graphical representation, and we exercise the double role of the axes explicitly at an early stage in the general presentation of the function concept. Thus exercise like problems 4 and 6 are now included in the student texts.

Difficulties in identifying images and (preimage, image) pairs for functions given in algebraic form. The following is Dafna's answer (grade 9) to problem 5.

a) 2   yes / no   2 is a natural number
   −1   yes / no   not natural number
    0   yes / no   not natural number
   11.5   yes / no   fractions are not natural
   1267   yes / no   natural number

b) −2   yes / no   not natural number
   10   yes / no   natural number
    8   yes / no   natural number
   46   yes / no   natural number
   23   yes / no   natural number

c) (5,26)   yes / no   both numbers are natural
   (0.5,8)   yes / no   0.5 is not a natural number
   (2,10)   yes / no   natural numbers

Typically, students correctly identified preimages in section (a), simply checking to see if the numbers belonged to the domain. But the correct procedure in sections (b) and (c) is more complex, and many students, like Dafna, came to the wrong conclusion because they ignored part of it. In order to determine if a given number is an image of the function f, three operations are required: (1) to check to see if the number belongs to the range, (2) to calculate the preimage, and (3) to check to see if this preimage belongs to the domain. Three similar operations are required to identify (preimage, image) pairs: to check to see (1) if the first number belongs to the domain, (2) if the second belongs to the range, and (3) if the second number is the image of the first under the given function. For example, in the pair (2, 10), 2 belongs to the domain and 10 belongs to the range, but $4 \times 2 + 6 = 10$. Clearly Dafna completed only one step in section (b) and two steps in section (c). Few students worked all three steps correctly. As we shall see again later, students have difficulty with tasks that involve a number of steps, ignoring one or more. Which step they ignore depends on the way the problem is posed. We shall return to this difficulty. Apparently, there is also often no distinction in the minds of the students between the image set and the range. This appears explicitly in the next difficulty.

Difficulties in distinguishing between the image set and the range. We noticed this difficulty in a number of problems. In problem 11, for example, students said that graph (c) does not represent a function of the type required by the question, apparently because the image set is only a subset of the given range.

Again, in the answer to problem 12, students gave very complex functions, just to "cover" all the natural numbers. For example, Ruth (grade 9) gave this answer:

$$f : \{\text{real numbers}\} \rightarrow \{\text{natural numbers}\}$$
$$f(x) = |\,[x]\,|$$

$(|\,[x]\,|$ is the absolute value of the largest integer not exceeding x.) Yet, a constant function with image a natural number would seem to be a much simpler example.

The difference between the image set and the range is, in our view, of sufficient importance to try to overcome this difficulty. This can be done by asking specific questions that require students to determine the image set and compare it with the range and to illustrate the two sets in specific cases—in set diagrams, in the coordinate system, and in set notation. Questions like those given in the examples here, being more complex, should be reserved for a more advanced stage.

Ignoring the domain and range of a function. We saw earlier that in questions involving more than one stage students ignored the rule of correspondence. Yet in other problems "the function is the rule," and they ignored domain and range. For example, in problem 8, which is presented in the same form as problem 5, students said that the function g given in section

(a) is equal to f. They also said that the function in section (c) is equal to f. Thus in both cases they ignored the domain and range. Again, in problem 9, many did not specify the domain and range of the function, although they were specifically asked to do so. In a problem of transfer from the algebraic form to the graphical (see fig. 5.3a) most of the students drew the answer as shown in figure 5.3b.

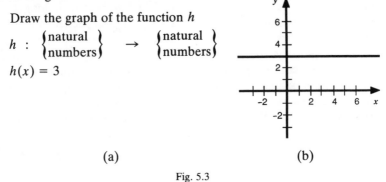

Draw the graph of the function h

$$h : \left\{ \begin{array}{l} \text{natural} \\ \text{numbers} \end{array} \right\} \rightarrow \left\{ \begin{array}{l} \text{natural} \\ \text{numbers} \end{array} \right\}$$

$$h(x) = 3$$

(a)   (b)

Fig. 5.3

Clearly some revision of the curriculum materials is required here, but we also have to consider whether it would not be advisable and justifiable to ignore some of these difficulties, as we shall explain. In order to "convince" the students that the function is influenced not only by the rule of correspondence but also by the domain, we suggest problems of the type shown in figure 5.4. This sort of question is also useful as a review of the different sets of numbers, which appeared as an incidental difficulty in our study.

Draw the graph of each of the following functions

$$f : \left\{ \begin{array}{l} \text{real} \\ \text{numbers} \end{array} \right\} \rightarrow \left\{ \begin{array}{l} \text{real} \\ \text{numbers} \end{array} \right\}$$
$$f(x) = 3$$

$$g : \left\{ \begin{array}{l} \text{positive} \\ \text{numbers} \end{array} \right\} \rightarrow \left\{ \begin{array}{l} \text{real} \\ \text{numbers} \end{array} \right\}$$
$$g(x) = 3$$

$$h : \left\{ \text{integers} \right\} \rightarrow \left\{ \begin{array}{l} \text{real} \\ \text{numbers} \end{array} \right\}$$
$$h(x) = 3$$

$$m : \left\{ \begin{array}{l} \text{natural} \\ \text{numbers} \end{array} \right\} \rightarrow \left\{ \begin{array}{l} \text{real} \\ \text{numbers} \end{array} \right\}$$
$$m(x) = 3$$

Fig. 5.4

However, the complexity of the function concept is also partly responsible for students' difficulties. We note that the definition of function, as now taught, involves many concepts—domain, range, image set, rule of correspondence. Either we have to make sure that these function concepts are understood in all representations before we continue teaching more about functions or we can choose to play down some aspects. Certainly with weaker students we would regard it as defensible to omit entirely any *explicit* treatment of equal functions as in problem 8 and not ask for the domain and range in problem 9. Also, sections (b) and (c) in problem 5 could be modified. For example:

b) (1) Find the preimages of 10 and 46.

(2) Explain why 8 and 23 cannot be images.

(*Or:* Is there an *x* in the domain for which $f(x) = 8$? Is there an *x* in the domain for which $f(x) = 23$?)

Misconceptions

Students often have the misconception that every function is a linear function (see also Markovits, Eylon, and Bruckheimer 1983). The following answers illustrate students' "linearity."

Ron (grade 9), problem 13

a)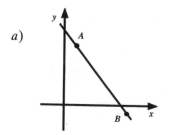

Iael (grade 9), problem 14

a)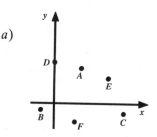

b) The number of different such functions that can be drawn is—

- 0
- ①
- 2
- more than 2 but fewer than 10
- more than 10 but not infinite
- infinite

Explain your answer: *Two points can be connected by only one straight line.*

b) The number of different such functions that can be drawn is—

- ⓪
- 1
- 2
- more than 2 but fewer than 10
- more than 10 but not infinite
- infinite

Explain your answer: *If I draw a function such that all the points are on it, what will happen is that for every x will be two y, and it will not be a function.*

Galia (grade 10), problem 15

a) $f(3) = 4$ $f(6) = 7$ $f(8) = 13$

$f(x) = x - 1$

$f(x) = x - 5$

b) The number of different such functions is—

- ⊙ 0
- 1
- 2
- more than 2 but fewer than 10
- more than 10 but not infinite
- infinite

In problems 13 and 14—and in similar problems given in graphical form—many students drew one straight line or a number of straight lines. When only one point was given, most of the students said correctly that the number of different functions is infinite (presumably because an infinite number of straight lines pass through one point). But when two or three points were given, the most common answer was one function only. For four or more points, the most frequent response was that no such function exists.

In the algebraic representation the students could somehow manage when one or two (preimage, image) pairs were given, but when a linear function did not suit, as in problem 15, only *one* student gave a correct answer.

Further, students gave linear examples when they were asked to give examples and nonexamples of functions (problem 3). Their conception of functions as linear would seem to be influenced by geometry (which they learn simultaneously with algebra in grade 9) and also by the time spent in the curriculum exclusively on linear functions. In order to change this conception, we suggest including nonlinear functions throughout the chapter dealing with linear functions as well as including problems on functions satisfying given constraints, such as problems 12–15. This latter type of problem caused trouble, and not only because the students' concept of function was mainly restricted to the linear. In our experience, the discussion of such problems with students results in a broader understanding of the function concept.

Difficulties

Students often have difficulties with certain types of functions.

The constant function. When the function is constant, all the preimages have the same image. The following fairly typical answers to problem 7*a* show that this was not understood.

Noa (grade 10)		Dan (grade 10)	
$g(4) = -4$	$g(-7) = 7$	$g(4) = -28$	$g(-7) = 49$
$g(0) = 0$	$g(3.5) = -3.5$	$g(0) = 0$	$g(3.5) = -24.5$

Difficulties with the constant function also occurred in a problem similar to problem 5, but with this function:

$$g : \begin{Bmatrix} \text{real} \\ \text{numbers} \end{Bmatrix} \rightarrow \begin{Bmatrix} \text{real} \\ \text{numbers} \end{Bmatrix}$$
$$g(x) = 4$$

The students were asked which of the numbers 12, 0, 4, 3.3 can be images under g. Very few students understood the question. Difficulties with the constant function were also found by Marnyanskii (1969). The students' incorrect responses suggest possibilities for corrective instruction. For example, the following:

Given the following three functions

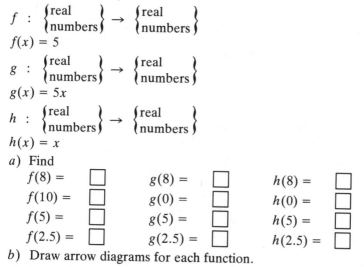

$$f : \begin{Bmatrix} \text{real} \\ \text{numbers} \end{Bmatrix} \rightarrow \begin{Bmatrix} \text{real} \\ \text{numbers} \end{Bmatrix}$$
$$f(x) = 5$$

$$g : \begin{Bmatrix} \text{real} \\ \text{numbers} \end{Bmatrix} \rightarrow \begin{Bmatrix} \text{real} \\ \text{numbers} \end{Bmatrix}$$
$$g(x) = 5x$$

$$h : \begin{Bmatrix} \text{real} \\ \text{numbers} \end{Bmatrix} \rightarrow \begin{Bmatrix} \text{real} \\ \text{numbers} \end{Bmatrix}$$
$$h(x) = x$$

a) Find

$f(8) = $ ☐	$g(8) = $ ☐	$h(8) = $ ☐
$f(10) = $ ☐	$g(0) = $ ☐	$h(0) = $ ☐
$f(5) = $ ☐	$g(5) = $ ☐	$h(5) = $ ☐
$f(2.5) = $ ☐	$g(2.5) = $ ☐	$h(2.5) = $ ☐

b) Draw arrow diagrams for each function.
c) Draw the graph of each function.

Functions represented by a disconnected graph. In problem 8 students said that the graph in section (d) is not a function, because "the points are not connected." In problem 1, they decided that the relation is not a function for the same reason.

This difficulty may not appear in the algebraic representation. Thus students will agree that the following is a function:

$$k : \begin{Bmatrix} \text{natural} \\ \text{numbers} \end{Bmatrix} \rightarrow \begin{Bmatrix} \text{natural} \\ \text{numbers} \end{Bmatrix}$$
$$k(x) = 4x$$

Problems like those we suggested in figure 5.4 should also help here.

Functions defined piecewise. The following answer was given by Noa (grade 10) to problem 2:

The relation is not a function, because every preimage has two images.

Students did not appreciate that the two rules of correspondence refer to two disjoint parts of the domain. They failed to understand that a function, in algebraic form, may be given by several rules of correspondence, each in some other part of the domain. This may also have contributed to their failure to give correct answers to such problems as number 15 and may have an influence on their concept of function being restricted to linear examples.

The concept of piecewise functions and disconnected graphs is of sufficient importance, especially to those continuing with mathematics and science in high school, to merit more attention in the curriculum. Again we can use the device of presenting questions of *apparent* similarity, and then use the dissimilarity to draw students' attention to the points of difficulty:

a) Draw the graphs of the following functions:

$$m : \begin{Bmatrix} \text{real} \\ \text{numbers} \end{Bmatrix} \to \begin{Bmatrix} \text{real} \\ \text{numbers} \end{Bmatrix}$$

$$m(x) = x + 1$$

$$h : \begin{Bmatrix} \text{real} \\ \text{numbers} \end{Bmatrix} \to \begin{Bmatrix} \text{real} \\ \text{numbers} \end{Bmatrix}$$

$$h(x) = \begin{cases} x + 1 & x \leq 5 \\ 6 & x > 5 \end{cases}$$

$$k : \begin{Bmatrix} \text{real} \\ \text{numbers} \end{Bmatrix} \to \begin{Bmatrix} \text{real} \\ \text{numbers} \end{Bmatrix}$$

$$k(x) = \begin{cases} x + 1 & x < 2 \\ 6 & x \geq 2 \end{cases}$$

b) Find:

$m(1) = \Box$	$h(1) = \Box$	$k(1) = \Box$
$m(2) = \Box$	$h(2) = \Box$	$k(2) = \Box$
$m(5) = \Box$	$h(5) = \Box$	$k(5) = \Box$
$m(10) = \Box$	$h(10) = \Box$	$k(10) = \Box$

It is interesting to note that such questions, although "obvious" now, did not suggest themselves to us in the original development of the curriculum materials. They arose as a direct response to our study of student difficulties.

Difficulties caused by technical manipulations. The complexity of the technical manipulations involved can cause difficulties in every topic. In the particular context of functions, we were not surprised to find that students had more difficulty in solving a task in which the rule of correspondence contained fractions, say, than a similar task involving integers only. But the types of tasks differ in their technical complexity as well. Thus the following problem and the responses given show (as might be expected) that when the function is given in algebraic form, it is more difficult to find preimages for given images than vice versa.

Amir (grade 10)

Given the function f

$$f : \quad \begin{array}{l} \text{real} \\ \text{numbers} \end{array} \rightarrow \begin{array}{l} \text{real} \\ \text{numbers} \end{array}$$

$$f(x) = -\frac{1}{2}x - 3$$

Find:

$f(4) = \boxed{-5}$ $\qquad$ $f(\boxed{2}) = -4$

$f(0) = \boxed{-3}$ $\qquad$ $f(\boxed{-1}) = 2$

$f\left(\frac{1}{2}\right) = \boxed{-3\frac{1}{4}}$ $\qquad$ $f(\boxed{}) = 0$

$f(-12) = \boxed{3}$

Amir, who found all four images correctly, found only one preimage correctly. The cause would seem to be clear: in order to find the preimages, the student has to solve an equation, whereas the images are found directly by substitution.

Another example of this category of difficulty occurs in the transfer from one representation of a function to another. We found that the transfer from algebra to graph was easier than the transfer from graph to algebra. This is most likely caused by the more complicated manipulations involved in the transfer from graph to algebra.

The difficulties caused by the complexity of technical manipulations were much more evident among students of lower ability than some of the other categories of difficulty we have discussed. Without crying "Back to basics," this is clearly an area that we cannot afford to neglect. One cannot solve problems without these technical manipulative skills. However, we should find ways of making the acquisition of these skills interesting.

CONCLUSION

Two further general points relevant to the curriculum materials became apparent during the study:

• We have evidence that it was easier for the students to manage functions given in graphical form than in algebraic form. It is not hard to find reasons for this. The graphical representation is more visual; the domain, range, and rule of correspondence are given together; and one obtains a visual impression of the behavior of the function. But in almost all curricula, the algebraic representation is taught before the graphical representation. We suggest that much more work in the early development of function concepts should be done in the graphical form.

• Our study also included tasks with a "story"—for example, the following bacteria task, which is mathematically similar to problem 13.

In a bacteria culture the number of bacteria is dependent on temperature. The number of bacteria per cm^3 at temperatures of 10°C and 20°C is marked in the coordinate system.

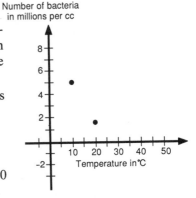

a) Draw a graph that you think describes the relationship between the number of bacteria and the temperature.

b) The different number of graphs that can be drawn is—
• 0
• 1
• 2
• more than 2 but fewer than 10
• more than 10 but not infinite
• infinite

Explain your choice: _____

Students of lower ability often found it easier to deal with tasks with a story than with tasks given mathematically. Thus it may be advisable to include such exercises more frequently in the curriculum materials.

We have described some difficulties and misconceptions that ninth- and tenth-grade students (who range in age from 14 to 16) have with the function concept. Our main object in undertaking the study was to provide a data base for future curriculum revision. In this sense, one of the most important outcomes of the study was the awareness of the specific difficulties because this awareness can result in suggestions as to how some of the difficulties can be removed or lessened.

REFERENCES

Markovits, Zvia, Bat Sheva Eylon, and Maxim Bruckheimer. "Functions: Linearity Unconstrained." In *Proceedings of the Seventh Conference of the International Group for the Psychology of Mathematics Education* (PME), pp. 271–77. Rehovot, Israel: Weizmann Institute, Department of Science Teaching, 1983.

Marnyanskii, I. A. "Psychological Characteristics of Pupils' Assimilation of the Concept of a Function." In *Problems of Instruction,* pp. 163–72. Soviet Studies in the Psychology of Learning and Teaching Mathematics, vol. 12. Reston, Va.: National Council of Teachers of Mathematics, 1969.

APPENDIX

Correct responses to the problems in figure 5.1

1. The relation is a function.

2. The relation f is a function.

4. Preimages: A, B, G on the x-axis (domain)
   Images: B, E on the y-axis (range)
   (Preimage, image) pairs: A, E, C on the graph
   Not (preimage, image) pairs: B, D, F, G

5. a) 2, 1267; the others are not natural numbers.
   b) 10, 46; the others are not images of natural numbers.
   c) (5,26) only. 0.5 is not a natural number and 10 is not the image of 2.

6.

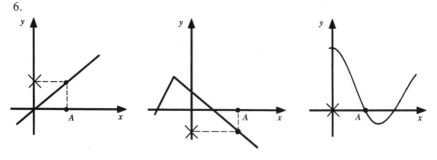

7. a) $g(4) = -7$ $g(-7) = -7$
       $g(0) = -7$ $g(3.5) = -7$
   b) No. The function is a constant function with the one image -7.
   c) Any real number can be substituted for x.

8. a) No. Both the domain and range are not the same as for the given function.
   b) No. The rule is different.
   c) No. The domain here is the set of positive real numbers.
   d) Yes. The domain and rule of correspondence are the same. We may also assume the range to be the same, but a student may be able to justify the answer "No" by reference to the range.

9. $f : \left\{ \begin{array}{l} \text{real} \\ \text{numbers} \end{array} \right\} \rightarrow \left\{ \begin{array}{l} \text{real} \\ \text{numbers} \end{array} \right\}$

   $f(x) = \begin{cases} 4 & x < 3 \\ 2 & x \geq 3 \end{cases}$

   The range can, in fact, be any set of numbers containing 2 and 4.

10.

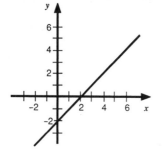

11. (a) and (c)

12. *a*) $g : \left\{{\text{real} \atop \text{numbers}}\right\} \rightarrow \left\{{\text{natural} \atop \text{numbers}}\right\}$

   $g(x) = 6$

   b) The number is infinite. For example, we can replace 6 by any real number.

13. *a*)

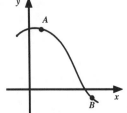

   b) The number is infinite. Any curve (which represents a function) and passes through *A* and *B* satisfies the conditions.

14. *a*)

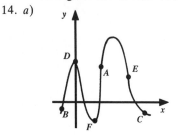

   b) The number is infinite. Any curve (which represents a function) and passes through the 6 points satisfies the conditions.

15. *a*) For example,

   $f : \left\{{\text{real} \atop \text{numbers}}\right\} \rightarrow \left\{{\text{real} \atop \text{numbers}}\right\}$ or $f : \left\{3,6,8\right\} \rightarrow \left\{4,7,13\right\}$

   $f(x) = \begin{cases} 4 & x \le 3 \\ 7 & 3 < x \le 8 \\ 13 & x > 8 \end{cases}$ $f(x) = \begin{cases} 4 & x = 3 \\ 7 & x = 6 \\ 13 & x = 8 \end{cases}$

   or $f : \left\{{\text{real} \atop \text{numbers}}\right\} \rightarrow \left\{{\text{real} \atop \text{numbers}}\right\}$

   $f(x) = \begin{matrix} x + 1 & x < 6 \\ x + 5 & x > 6 \end{matrix}$

   b) The number is infinite. For example, $f : \{ \text{real numbers} \} \rightarrow \{ \text{real numbers} \}$

   $f(x) = \begin{cases} a & x < 3 \\ 4 & x = 3 \\ a & 3 < x < 6 \\ 7 & x = 6 \\ a & 6 < x < 8 \\ 13 & x = 8 \\ a & x > 8 \end{cases}$

   where *a* is any real number, gives an infinite set of functions.

6

Establishing Fundamental Concepts through Numerical Problem Solving

Franklin Demana
Joan Leitzel

F OR many students the first algebra course is full of mysteries that they never fully understand. The second algebra course is then only marginally mastered by these students, who must depend heavily on imitation and repetition for success. This paper will demonstrate that students can understand basic concepts of algebra when they are introduced through numerical computation and problem solving before they are encountered in more formal courses in algebra. The first algebra course is not mysterious for students who have such understanding before they begin.

The approach to algebra described here is the same as that used in a four-year project in Ohio that gave attention to the need for students to work with key algebra concepts in a numerical setting before they confronted the concepts in formal courses in algebra. (The project was Approaching Algebra Numerically [AAN], supported by Standard Oil of Ohio, 1984–85, and by the National Science Foundation, 1985–87.)

This approach depends heavily on calculators and problem solving. In order to strengthen students' understanding of arithmetic concepts that are basic to algebra, they are first led to investigate how calculators work. Once

61

they are comfortable with calculators, they solve problems numerically by building tables and using guess-and-check procedures. Next, students revisit the same problem situations, but this time they investigate the problems geometrically by making graphs of the relationships in the problems. Finally, they revisit the problem situations once again and use tables to support writing equations that represent the problems. Then they solve the problems by solving the equations.

USING COMPUTATION TO ANTICIPATE ALGEBRA

Order of Operations

Learning to use a calculator with algebraic logic and hierarchy requires students to understand basic arithmetic properties that are important in algebra. For example, understanding the order of operations is essential in evaluating algebraic expressions like the following:

Algebra. Find the value of $7 + 3x^2$ when x is 0.5. To be ready for this exercise students should do analogous types of numerical computation in prealgebra:

Prealgebra. Find the value of $7 + 3 \times 0.5^2$. The important thing for students to understand here is that in mathematical hierarchy, exponentiation is performed first, then multiplication, and finally addition. Using a calculator with the same hierarchy as the usual order of operations reinforces this understanding.

The correct use of parentheses is another technical feature of computation that is essential in algebra.

Algebra. Does $3(x + 2)$ equal $3x + 2$? This example suggests a common mistake of beginning algebra students. To avoid this mistake, prealgebra students need computational experiences like the following:

Prealgebra. In the keying sequence

$$\boxed{(}\ \boxed{4}\ \boxed{+}\ \boxed{7}\ \boxed{)}\ \boxed{\times}\ \boxed{3}\ \boxed{-}\ \boxed{6}\ \boxed{=},$$

what does the calculator do when the $\boxed{)}$ key is pressed? When the $\boxed{-}$ key is pressed? The calculator adds 4 and 7 when the $\boxed{)}$ key is pressed and then multiplies the result by 3 when the $\boxed{-}$ key is pressed. This example helps students see that parentheses are needed to interrupt the usual order of operations.

Prealgebra. Does omitting parentheses change the value of $4(8 + 3)$? Students using calculators with hierarchy will find that the value of $4(8 + 3)$ is 44 and the value of $4 \times 8 + 3$ is 35, so parentheses cannot be omitted without changing the value of the expression. It should be noted that the teacher is now in a position to focus attention on the distributive property by having students compare the values of $4(8 + 3)$ and $4 \times 8 + 4 \times 3$.

Negative Numbers

Students need to be comfortable with negative numbers before they begin algebra. Calculators are particularly effective in gaining this experience. Fortunately the sign change key is different from the subtraction key. And signs of powers of negative numbers are visible on calculators. Students can compare powers of positive numbers and powers of negative numbers.

$$3 = 3 \qquad\qquad -3 = -3$$
$$3^2 = 9 \qquad\qquad (-3)^2 = 9$$
$$3^3 = 27 \qquad\qquad (-3)^3 = -27$$
$$3^4 = 81 \qquad\qquad (-3)^4 = 81$$

If these lists are extended, not only do students see that $(-3)^7 = -3^7$, but they know this equality means that raising the opposite of 3 to the seventh power is the same as first raising 3 to the seventh power and then taking the opposite. This understanding is essential to evaluating polynomial expressions for negative values of the variable.

Algebra. Find the value of x^{17} when x is -3.

Prealgebra. Find each negative number in this list of numbers:

$$(-3)^2, (-3)^3, (-3)^{42}, (-3)^{55}$$

Other Features

There are other features of calculators that make them particularly effective in prealgebra instruction. Calculators reinforce the fact that a fraction is a quotient, that a fraction bar is a grouping symbol, and that we can't divide by 0. Students who use calculators gain computational experience with exponentiation, with square root, and with scientific notation, all at a much deeper level than they could experience without computational aids. This experience serves students well in algebra.

INTRODUCING VARIABLES

Calculators permit students to investigate rich problem situations and to compute relationships in a problem situation in many special cases. They are able to generalize from these cases as a consequence of the computation. Expressing generalized relationships is basic to algebra. However, beginning students typically do not see a need to generalize. They do not even see that generalization is natural, because usually only one question is asked in each problem setting they confront. Calculators make investigating many cases possible, make generalizing meaningful, and make it very natural to use a variable as a tool in expressing a generalization.

Figure 6.1 gives a simple example of a problem setting appropriate for investigation by middle school students.

Damon's Department Store marks up wholesale prices 35% to determine retail prices. Complete the following table:

Wholesale Price ($)	Retail Price($)
5.00	$5.00 + 5.00 \times .35 = 6.75$
7.98	$7.98 + 7.98 \times .35 = 10.77$
15.40	$15.40 + 15.40 \times .35 = 20.79$
20.00	$20.00 + 20.00 \times .35 = 27.00$
32.00	43.20
P	$P + P \times .35$

Fig. 6.1

Students gain experience in several areas from this problem. They understand that the retail price is computed from the wholesale price, and without formal language or notation they compute with a function. From their computation, they understand the pattern for describing retail price in terms of wholesale price. At first this pattern should be described verbally; in time, students can use a variable to write the retail price in terms of the wholesale price.

To answer the question, "What is the wholesale price of an item whose retail price is $43.20?" students with calculators use guess-and-check procedures. Usually they become quite quick in refining estimates and getting answers, thus handling such questions numerically long before they are ready to write and solve equations to solve such problems.

Before students are introduced to algebraic methods, it is useful for them to display graphically the relationships in a problem like the one in figure 6.1. The numerical table provides information for the first few points of the graph. Other points can be computed until the complete graph is clear. Then many questions of the type "What was the wholesale price, if the retail price is . . . ?" can be answered by reading the graph. The concept of function can be visualized as a graph—a far more concrete representation for a function than an algebraic expression.

UNDERSTANDING VARIABLES

It is well known that students have difficulty with the concept of variable and that this difficulty can be basic to a lack of success in algebra. We have found that introducing variables to represent functional relationships in concrete problem situations gives students the view that variables can represent numbers from large sets of numbers and that variables are useful tools in describing generalizations. When variables are introduced in tables like the one in figure 6.1, students see that all the numerical information in the table is summarized in the last line. The variable is a powerful tool to express all the special cases in a concise manner.

SIMPLIFYING ALGEBRAIC EXPRESSIONS

When variables are introduced in tables to express generalized relationships, students gain practice in writing algebraic expressions. In the previous example, a student might write $P + P \times .35$ to express the retail price in terms of the wholesale price P. Initially, we do not require that students conform to all the conventions of algebraic notation. In time, they are willing to write constants in front of variables and drop the multiplication symbol, writing $P + .35P$ instead of $P + P \times .35$. However, we have observed that such conventions are barriers to understanding for many students and cannot be assumed quickly.

The introduction to algebraic expressions through problem solving and computation makes simplifying expressions useful and natural. In order to simplify $P + .35P$, students confront the distributive property, write expressions in factored form, and combine like terms:

$$P + .35P = (1 + .35)P = 1.35P$$

Although students may be able to succeed in arithmetic without understanding the distributive property, that understanding is essential in algebra. Students must be able to represent expressions in both factored and expanded forms in order to simplify expressions. In prealgebra we look for opportunities for students to get this practice in concrete problem situations such as the one in figure 6.2.

For some rectangles, the length of the rectangle is four centimeters more than the width. Complete the following table:

Width (cm)	Length (cm)	Perimeter (cm)	Area (cm²)
1	5	12	5
5	9	28	45
8.4	12.4	41.6	104.16
w	$w + 4$	$4w + 8$	$w^2 + 4w$

Fig. 6.2

For the perimeter entry in the last line of this table, different students write the expressions $2(w + w + 4)$, $2w + 2(w + 4)$, and $4w + 8$. Within this numerical setting, they appreciate that an expression has several forms. They are motivated to simplify the more complicated expressions when they see how much easier it is to compute perimeter using the expression $4w + 8$ rather than $2(w + w + 4)$ or $2w + 2(w + 4)$.

WRITING EQUATIONS AND FINDING SOLUTIONS

Tables that contain numerical data about a problem situation and also contain a variable to describe the general case provide an excellent opportunity for introducing prealgebra students to equations and solutions to equations. Using the completed table in the previous example, we can ask questions like these:

$$\text{Find } w \text{ if } w^2 + 4w = 45.$$
$$\text{Find } w \text{ if } 4w + 8 = 41.6.$$

Students see that the table contains the infomation they need to solve these equations. Furthermore, they can write equations and solve them to answer questions like "What is the width of one of the rectangles in this example if the perimeter is 28? If the area is 104.16?" In this way students learn both to represent problem questions as equations and to identify solutions to equations before they begin to learn algorithms for solving equations.

If students are not shown formal algorithms, they solve equations numerically, using tables that they build. Because they depend on computation to find solutions, they are motivated to simplify algebraic expressions before they begin to compute (fig. 6.3).

Find a solution to the equation $3x + 2(4x - 6) = 21$.

Student Work:

$3x + 2(4x - 6) = 21$	x	$11x - 12$
$3x + 8x - 12 = 21$	5	43
$11x - 12 = 21$	3	21

Solution: $x = 3$

Fig. 6.3

In time, students tend not to build tables to solve linear equations when one side of the equation is a constant. They "unpack" the variable by using their computational experience to reverse operations. Then they can write a solution. This is the way students might reason their way to a solution to the equation $11x - 12 = 21$:

Student Analysis: This is what I know about the solution. If I multiply it by 11 and then subtract 12 from the result, I get 21. Thus to find a solution, start with 21, add 12, and divide by 11.

Solution: $(21 + 12) \div 11$

Similar reasoning works for an equation as complex as $\dfrac{22x - 15}{.05} = 1592$.

Student Analysis: This is what I know about the solution. If I multiply it by 22 and subtract 15, then divide the result by .05, I get 1592. Thus to find a solution, start with 1592, multiply by .05, add 15, and divide by 22.

Solution: $(1592 \times .05 + 15) \div 22$

Some might argue that this unconventional way of identifying solutions to equations could interfere with students' later learning of conventional algorithms, like "do the same to both sides of the equation." However, if students' reasoning flows from strong numerical experience, it is good prealgebra. They should be told that in algebra they learn techniques to solve a few kinds of equations mechanically, but only a few kinds. Numerical methods may be the best they have for all the rest.

A variety of types of equations must be used in prealgebra—exponential equations, higher-order polynomial equations, equations with radicals—in order to anticipate later experiences in mathematics. Interesting problem situations that embody many types of functions are sought.

Example: Jennifer invests $500 at 7.5% compounded monthly. How long will it take for her investment to double?

A less experienced student in prealgebra will solve this problem by building a table. Then using guess-and-check procedures, the student will find an answer to the question. A prealgebra student who has already done considerable computation with compound interest will know that the expression

$$500 (1 + \frac{.075}{12})^N$$

describes the money in the account after N months. This student can write the equation

$$500 (1 + \frac{.075}{12})^N = 1000.$$

For a student with a calculator, solving this equation takes no longer than solving a linear equation. A student accustomed to reasoning numerically will observe that the solution is the value of N that makes $(1 + .075/12)^N$ approximately 2. On a calculator that number is quickly found.

CONCLUSION

Test results from the AAN project confirm that basic concepts of algebra are accessible to students in their arithmetic experiences through numerical computation and problem solving. In particular, the ideas of variable and function can be established, laying the foundation for success in the first algebra course and, as a consequence, in later college preparatory courses.

BIBLIOGRAPHY

Booth, Lesley R. *Algebra: Children's Strategies and Errors.* New Windsor, Berkshire, England: NFER-Nelson Publishing Co., 1984.

Collis, Kevin F. "Operational Thinking in Elementary Mathematics." In *Cognitive Development,* edited by John A. Keats, Kevin F. Collis, and Graeme S. Halford. New York: John Wiley & Sons, 1978.

Demana, Franklin D., Joan R. Leitzel, and Alan Osborne. *Seventh and Eighth Grade Units—Approaching Algebra Numerically Project.* Student Materials and Teacher Materials. Lexington, Mass.: D. C. Heath & Co., forthcoming.

Hart, Kathleen. *Secondary School Children's Understanding of Mathematics.* London: Chelsea College, Centre for Science and Mathematics Education Research Monographs, 1980.

Hart, K., and Lesley R. Booth. "Children Find Mathematics Difficult: The Results of the CSMS Research." *Bulletin of the Institute of Mathematics and Its Applications* 17 (May 1981): 114–15.

Küchemann, Dietmar E. "Children's Understanding of Numerical Variables." *Mathematics in School* 7 (September 1978): 23–26.

Leitzel, Joan R., and Alan Osborne. "Mathematical Alternatives for College Preparatory Students." In *The Secondary School Mathematics Curriculum,* 1985 Yearbook of the National Council of Teachers of Mathematics, pp. 150–65. Reston, Va.: The Council, 1985.

McKnight, Curtis C., Kenneth J. Travers, and John A. Dossey. "Twelfth-Grade Mathematics in U.S. High Schools: A Report from the Second International Mathematics Study." *Mathematics Teacher* 78 (April 1985): 292–300.

Travers, Kenneth J. *Second International Mathematics Study, Detailed National Report United States.* Champaign, Ill.: Stipes Publishing Co., 1985.

CAN YOUR ALGEBRA CLASS SOLVE THIS?

Problem 5. The sum of two numbers is 28. The product of the numbers is 7. Find the sum of the reciprocals of the numbers.

Solution on page 248

7

Algebraic Instruction for the Younger Child

Frances M. Thompson

DURING the past decade many researchers have tested, clarified, or modified the developmental learning theories of Piaget. One major conclusion must be drawn from these studies: each child learns at his or her own developmental pace, progressing sequentially from the concrete operational level to the abstract level.

For each new concept studied, problems are initially "solved" by observing what happens as concrete objects are manipulated. Gradually, as the concrete actions are assimilated, the learner is able to reenact the earlier concrete actions by using pictures that represent the objects. Only after further intellectual development is the learner able to reproduce these actions by means of abstract symbols and operations.

This sequence of learning must be used for each new major concept introduced to elementary school children. The amount of time needed to progress from the concrete level to the abstract level, however, depends on both the concept to be studied and the mental maturity of the child. The primary student may take several weeks to learn a given concept, whereas the intermediate student may acquire a concept in one or two days.

In my recent research with students of varying abilities in grades 3 through 6, I have developed and implemented several instructional sequences based on the learning phases described above. This has been done in an effort to determine what particular *algebraic* concepts elementary school children might be able to comprehend and what instructional approaches might be appropriate. The particular concepts taught have been the integers and their operations and simple linear equations such as $3B - 5 = -2$.

CONCRETE REPRESENTATION

In each sequence a concept is introduced with concrete materials first, then reviewed with pictorial models representing the concrete objects. As each concrete or pictorial step is completed by the students, the action is recorded on the chalkboard or on student papers by means of abstract notation. A packet of red and blue one-inch square pieces of construction paper is given to each student. The blue squares are marked with an X on each side in order to facilitate the later pictorial work in which color is not used. The red squares are not marked. A flannel board with felt pieces similar to, but larger than, the children's squares is also used for class demonstrations by the teacher and students.

The red and blue pieces are introduced as "magic checkers"—magic in that when a red and a blue checker are paired, they both "disappear." The children quickly recognize that such a combination acts just like "zero," a concept already familiar to them. A single red checker is named "one" and written as "1," and a single blue checker is named the "inverse, or opposite, of one," the "inverse of a red checker," or the "negative of one" and written as "1⊗." Thus, "5⊗" represents "five negative ones," "negative five," or the "inverse, or opposite, of five."

The different names for a symbol are used interchangeably during all discussion sessions. The use of ⊗ in the notation also corresponds to the actual marking on the blue squares⊠ , thereby making it easier for the children to bridge from the concrete material to the symbols. The children are very comfortable with the notation and labeling, working easily and simultaneously with the oral and written language forms. (The symbol -5 has been avoided in the research efforts thus far in order to prevent possible confusion with the subtraction sign.)

Several examples of forming zero-pairs with inverses are first worked on the flannel board by the teacher or a student. As an example, if four red pieces and three blue pieces are on the board, a student matches each blue piece with a red one and removes each pair formed, leaving a red piece (see fig. 7.1). After the pairing action is completed, the student orally describes the problem as "four ones and three inverse ones are the same as one." Then this statement is immediately recorded by teacher or student on the chalkboard as $4 + 3⊗ = 1$. This, of course, introduces the class to the

Fig. 7.1

addition of integers. The students work other problems at their desks with their own paper "checkers" and share their answers with the rest of the class.

PICTORIAL REPRESENTATION

Once the students have practiced sufficiently with the concrete materials, addition problems are presented on special worksheets that use drawings closely resembling the felt and paper squares. Only the red and blue colors are missing. Sometimes the two numerical addends are given and students have to draw the needed squares; in other problems, the squares for the addends are furnished and students have to record the correct numbers being added. Students form zero-pairs by using connecting curves. The remaining unmatched squares represent the answer to the problem, which the students record in the blank provided on the worksheet (see fig. 7.2).

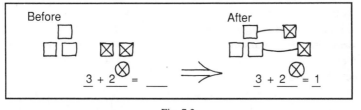

Fig. 7.2

INTRODUCING SUBTRACTION

Subtraction is introduced as "take away," a familiar approach to elementary students. Problems like "If there are five red pieces and two red pieces are taken away, then there are three red pieces left" are first used to introduce the concept. This result is recorded on the chalkboard as $5 - 2 = 3$. Next comes the problem, "If there are three blue pieces or inverses and one inverse piece is removed, then there are two inverse pieces left." This is recorded as $3^{\oplus} - 1^{\oplus} = 2^{\oplus}$.

Now for a "big" problem! "What if there are three red squares and we want to take away two blue ones? Since there are no blue pieces, how can we get some without changing the value of the original set of squares?" Students typically volunteer to add zero-pairs to the set in order to obtain the needed blue pieces. (See fig. 7.3.) When two red-blue pairs of felt pieces are placed

Fig. 7.3

on the flannel board, the result is five red pieces and two blue pieces on the board. The desired two blue pieces are then removed, leaving five red ones. This is recorded as $3 - 2^{\oplus} = 5$. The sentence is read as "three minus two inverses is five."

Students readily observe that taking the inverse of 2 away from 3 has the same effect as *adding* 2 to 3, since bringing in two inverse ones also brings in two extra *ones*. Although the two inverse ones are later removed, the two ones remain. This observation is intentionally left at the intuitive level; no effort is made to make a "rule" out of it for the class.

When most students appear proficient in subtracting with concrete materials, they are given worksheets of subtraction problems in pictorial form. They have to draw any needed zero-pairs, then cross out the amount to be subtracted, leaving the answer. This procedure supports the idea observed at the concrete level that subtracting a number is equivalent to adding its opposite or inverse form. The numbers involved are recorded in number sentence form. An example is shown in figure 7.4.

As the pictorial worksheets are being completed at the students' desks, problems are also demonstrated with the felt squares on the flannel board, and the corresponding "pictures" of the problems are drawn on the chalkboard. On both the addition and subtraction pictorial worksheets, the

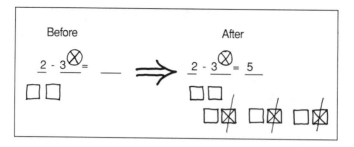

Fig. 7.4

last few problems are given only in number sentence form (e.g., 2 + 5⊗ = ___ or 3⊗ − 5 = ___). The students are told that they can solve the problems by drawing the necessary squares, by using the paper pieces, or by working the problems in their heads. This is done for each concept in an effort to learn which level of operation (concrete, pictorial, or abstract) individual children might prefer.

It has been observed that about 20 percent of the third-grade children choose to continue with the concrete materials as they simultaneously work with the pictorial models. The other third graders choose to work with the pictorial models alone. In the operation of addition, a few even try to match their drawings mentally.

At the fourth-, fifth-, and sixth-grade levels, most students initially choose to draw the necessary squares and match them. About 20 percent of the students at each of these three grade levels, however, record the correct answers without making any drawings. Their explanations of the problems indicate that they form mental images of the squares and match them in order to get their abstract answers.

INTRODUCING LINEAR EQUATIONS

The instructional sequencing above for addition and subtraction, both concrete and pictorial, usually requires about one hour to complete with third graders and about thirty minutes for sixth graders. When all students have completed their study of integer addition and subtraction satisfactorily, the idea of a variable in a *linear equation* can be introduced. Multiplication and division of integers are not necessary at this point as long as small whole numbers are used for the variable coefficients.

In order to represent an equation or equate two algebraic expressions concretely, a long piece of yarn is taped to hang vertically at the center of the flannel board to serve as the equals sign. The teacher places some red or blue checkers to the right of the yarn. A brown sheet of paper is then placed to the left of the yarn with an identical set of checkers hidden underneath it. (See fig. 7.5.) The students are asked to predict the value of the hidden checkers. The "hidden checkers" approach, although seemingly simple, successfully introduces the young student to the role of the variable in mathematics.

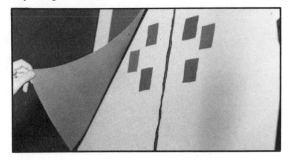

Fig. 7.5

After several different problems are discussed, the large brown paper is replaced on the flannel board by a smaller two-by-five-inch brown felt rectangle or bar, for convenience. When the brown bar (now orally named the "variable B") is placed on one side of the yarn and some red or blue checkers—for example, four blue checkers—are placed on the other side, the equation $B = -4$ is represented. Students soon realize that when the variable, or brown bar, is by itself on one side, the other side of the yarn must reveal the value of the variable.

With the role of the variable now established, the students are ready to solve some simple linear equations. As an example, two blue felt pieces are placed on the left side of the yarn along with the brown bar, and one red piece is placed on the right (see fig. 7.6). The students are asked to name the simplest value in checkers that the brown bar must have so that when these

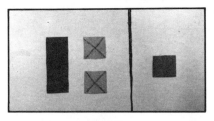

Fig. 7.6

checkers are matched with the two blue ones, the resulting value is one (i.e., the value of the red checker to the right of the yarn). The answer, three red checkers, is suggested by a student. To verify the answer, the brown bar is replaced by three red pieces, and all possible zero-pairs are formed. One red piece finally remains on the left side of the board, which matches the checker on the right.

Students soon discover a new stragegy: Adding two red checkers to the left side creates two zero-pairs and leaves the brown bar by itself. But as the children already know from earlier arithmetic experiences, any adjustment to one side of an equation or number sentence requires equivalent adjustment to the other side as well. Therefore, two red checkers must be added to the right side also. After forming and removing any possible zero-pairs on either side, the brown bar remains alone on the left and three red checkers remain on the right. This arrangement reveals the value of the bar (in this example, three red checkers), which corresponds to the number of checkers predicted and verified earlier.

The students now apply this new strategy to other problems with the paper checkers at their desks. One-by-three-inch brown paper rectangles symbolize the brown bar on the flannel board and serve as their variables. Pieces of yarn are used to separate their desk surface areas into two parts, thereby acting concretely as the equals sign. Problems are demonstrated on the flannel board by individual students, who must carefully explain their steps by properly using the terms *variable* and *inverse* for the brown bar and blue checker, respectively. Continuing to use the letter *B* for the brown bar, or variable, the students also record their steps on the chalkboard in the format shown in figure 7.7.

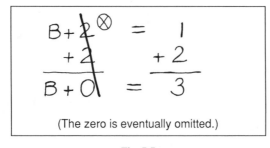

(The zero is eventually omitted.)

Fig. 7.7

This concrete method of introducing linear equations takes about one hour to develop with a class of third-grade students and about thirty minutes with fifth or sixth graders. After the concrete actions are mastered, students begin to solve equations using the pictorial model and, finally, the abstract model only. An example is shown in figure 7.8. Even problems with negative integers for answers, as in figure 7.8, do not seem to bother the children.

The third graders find it easy and enjoyable to solve linear equations on the flannel board or with the paper pieces at their desks (concrete level); however, solving with picture models or symbols seems more difficult. A few of the students at each grade level studied have been able to use the pictorial and abstract models successfully without the concrete materials, but no effort has been made thus far in the research studies to have all students

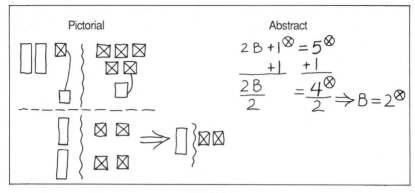

Fig. 7.8

solve problems at these two higher levels of learning. Most of the students still seem to be operating at the concrete level with respect to the concept of linear equations.

STUDENT REACTIONS

The algebraic terminology and symbolic notation for the concepts discussed above are easily understood by elementary students, particularly when they can associate the new terms and symbols with physical materials and actions. The students frequently interchange the new word labels with the actual physical descriptions of the objects being labeled. For example, sometimes they might say "the variable B," and at other times they might say "the brown bar." But the role the object plays mathematically always seems clearly understood. This language flexibility is quite common in young children and is mathematically and pedagogically acceptable as long as the proper abstract notation is used to record any concrete or pictorial actions that may occur.

Unlike the third and fourth graders, some fifth and sixth graders in several sessions appeared reluctant to interact with the concrete materials, perhaps from lack of experience with such materials or from a fear of appearing childish. Yet these students could not operate with integers mentally or abstractly either. After being encouraged individually, they began to work with the materials and were successful in solving their problems.

Sixth graders in an accelerated mathematics class had already been introduced to the abstract "rules" of integer addition and subtraction before a special lesson with the checkers. At first they wanted to find their answers by simply applying the rules. They soon realized that they could not explain mathematically what was happening in the solution process and began using the checkers to solve their problems. In further classroom studies with integers, these students continued to rely on these past concrete experiences with the checkers to verify their abstract algebraic procedures and solutions.

EXTENSIONS

Other lessons in algebra have also been presented to fourth, fifth, and sixth graders. Multiplication and division of integers have been introduced concretely, using the same colored squares described in this article. Concepts such as the addition and subtraction of first- and second-degree polynomials in one variable, simultaneous linear equations, linear inequalities, and binomial multiplication can also be demonstrated with concrete and pictorial models. These concepts need to be explored further with students at these grade levels.

Wide use of the instructional technique described above reveals that students in grades 3–6 can successfully learn simple algebraic concepts and are eager to do so if they are first allowed to operate on concrete materials. The only prerequisites needed are the four basic operations with whole numbers. It is exciting to think of all the mathematics young children may be capable of learning if they can just be taught through an instructional sequence according to their own developmental needs.

CAN YOUR ALGEBRA CLASS SOLVE THIS?

Problem 6. For a given arithmetic series, the sum of the first 50 terms is 200, and the sum of the next 50 terms is 2700. What is the first term of the series?

Solution on page 248

CAN YOUR ALGEBRA CLASS SOLVE THIS?

Problem 7. Find the difference between the larger root and the smaller root of

$$x^2 - px + \frac{(p^2 - 1)}{4} = 0.$$

Solution on page 248

8

Proportionality and the Development of Prealgebra Understandings

Thomas R. Post
Merlyn J. Behr
Richard Lesh

PROPORTIONAL reasoning is generally regarded as one of the components of formal thought acquired in adolescence. Relatively few junior high school students of average ability use proportional reasoning in a consistent fashion unaccompanied by physical actions (Goodstein 1983; Kurtz and Karplus 1979; Lawson and Wollman 1976; Lunzer 1965; Noelting 1980; Wollman and Karplus 1974). The issues involved in the teaching and learning of proportional reasoning are seemingly more complex than previously acknowledged. Repeatedly disappointing achievement results on the National Assessment of Educational Progress and international achievement comparisons further substantiate this perspective.

The majority of past attempts to define proportional reasoning (e.g., Karplus, Pulas, and Stage 1983; Noelting 1980) have been primarily concerned with individual responses to missing-value problems where three of four values in two rate pairs were given and the fourth was to be found. Those students who were able to answer successfully the numerically "awkward" situations containing noninteger multiples within and between the rate pairs were thought to be at the highest level and were considered proportional reasoners. We believe that this is a limited perspective, a necessary but not a sufficient condition, especially since these problems lend themselves to purely algorithmic solutions.

The development of this paper was supported in part by the National Science Foundation under Grant No. DPE-8470077 (The Rational Number Project). Any opinions, findings, and conclusions expressed are those of the authors and do not necessarily reflect the views of the National Science Foundation.

This paper attempts to expand the previous view and suggests that proportional reasoning encompasses a wider and more complex spectrum of cognitive abilities. As more data are analyzed and additional research conducted, we and others will surely continue to modify and improve on our understanding of this important construct.

WHAT IS PROPORTIONAL REASONING?

Proportional reasoning is one form of mathematical reasoning. It involves a sense of covariation, multiple comparisons, and the ability to mentally store and process several pieces of information. Proportional reasoning is very much concerned with inference and prediction and involves both qualitative and quantitative methods of thought. The fact that many aspects of our world operate according to proportional rules makes proportional reasoning abilities extremely useful in the interpretation of real-world phenomena.

Proportional reasoning has aspects that are both mathematical and psychological. Mathematically, all proportional relationships can be represented by the function $y = mx$, the most fundamental type of linear equation. This equation represents a simple relationship between ordered pairs of numbers (x, y) that is multiplicative in nature. Traditionally, proportional situations have been embedded in missing-value problems. ($a/b = c/x$, with a, b, and c usually given explicit values. The task is to determine the value of x; the position of x can vary.)

Proportional reasoning is also required to compare two given rate pairs. In numerical comparison situations one is asked to compare a/b and c/d where a, b, c, and d are all given. The task is to deduce which rate pair is greater, faster, darker, more expensive, more dense, and so on.

Proportional reasoning involves qualitative thinking: "Does this answer make sense? Should it be larger or smaller?" Such thinking requires a comparison that is not dependent on specific values. For example, "If Nicki ran fewer laps in more time that she did yesterday, would her running speed be faster, slower, the same, or can't tell?" (How about fewer laps in less time?) In this kind of situation, qualitative reasoning requires the ability to interpret the meaning of two ratios, store that information, and then compare these interpretations according to some predetermined criteria. This process requires a mental capability that Piaget has equated with the formal operational level of cognitive development. He referred to this process as operating on operations. That is, the interpretation of each of the ratios is an operation in and of itself, and the comparison is yet another level of operations. Such processing requires multilevel comparative thinking quite different from an algorithmic approach, where a rule is used to solve predictable problems in predetermined ways.

In one sense qualitative reasoning is more general than quantitative reasoning, since one's conclusions relate to an entire class of values rather than specific entities. In another sense, qualitative reasoning is an important means to check the feasibility of responses and a way to establish broad parameters for problem conditions. It is well known that experts in a wide variety of areas use qualitative approaches to problems as a means to better understand the situation before proceeding to actual calculations and the generation of an answer. Novices, however, tend to proceed directly to a calculation or a formula without the benefit of prior qualitative analyses. It should also be pointed out that novices often answer problems incorrectly, suggesting that they could benefit from the use of qualitative procedures.

Another aspect of proportional reasoning involves a firm grasp of various rational number concepts such as order and equivalence, the relationship between the unit and its parts, the meaning and interpretation of ratio, and issues dealing with division, especially as this relates to dividing smaller numbers by larger ones. A proportional reasoner has the mental flexibility to approach problems from multiple perspectives and at the same time has understandings that are stable enough not to be radically affected by large or "awkward" numbers or by the context within which a problem is posed.

And lastly, proportional reasoners must be able to distinguish between proportional and nonproportional situations. This has direct implications for instruction.

WHY IS PROPORTIONAL REASONING IMPORTANT TO THE LEARNING OF ALGEBRA?

1. Proportionality is a simple yet powerful example of a mathematical function and can be represented as a linear equation. As such, it is a convenient and perhaps necessary bridge between common numerical experiences and patterns and the more abstract relationships that will be expressed in algebraic form. The algebraic representation of proportionality ($y = mx$) represents an incredibly large class of physical occurrences.

2. Proportions (expressed as two equivalent ratios) are useful in a wide variety of problem-solving situations, such as the many types of rate problems—speed, mixture, density, scaling, conversion, consumption, pricing, and other types of comparisons are examples. An example of a speed-related rate is as follows: If a faucet drips eleven times in twenty seconds, how many drips are there in an hour?

Percent is a special type of rate. In percent situations the denominator of one rate pair will always be 100. Mathematics curricula in the past distinguished among the "three cases of percent." This is no longer done, since all percent-related situations can be solved with the use of proportions and an essentially identical conceptual framework. For example: (1) Jessica scored

85 points on a 115-point test. What percent was this? $(85/115 = x/100)$; (2) If Jessica scored 74% on a test with 115 items, how many did she get correct? $(74/100 = x/115)$; or (3) Jessica had 85 items on a test correct. This was 74%. How many items were on the test? $(74/100 = 85/x$ or $85/74 = x/100)$

3. Algebraic thought and understanding often involve different modes of representation. Tables, graphs, symbols (equations), pictures, and diagrams are all important ways in which algebraic ideas can be represented. The ability to generate and understand translations within and between these modes is an essential element of mathematical competence in all areas, not just algebra. Proportional situations and the reasoning that accompanies them provide an excellent vehicle within which to illustrate these multimodal associations. For example, a student given a table expressing a numerical relationship between two domains or measure spaces could be asked to construct an equation that defines the relationship. Of course these graphs, tables, pictures, and equations could (and should) occur in many different orders, with emphasis being placed on the translation process and the explanations of appropriate connections.

THE STANDARD ALGORITHM

Although it can be effectively argued that students need to automatize certain commonly used mathematical processes (Gagné 1983), it can likewise be argued that the most efficient methods are often those that are the least meaningful and therefore are to be avoided during the initial phases of instruction. Unfortunately we sometimes confuse efficiency and meaning, and by default, even with the best of intentions, we introduce a concept in the most efficient but least meaningful manner. The standard algorithm for proportionality—$a/b = c/x$, a, b, and c given, find x—is one of these areas. The standard solution procedure is to cross multiply and solve for x. That is, $ax = cb$, or $x = cb/a$.

This algorithm in and of itself is a mechanical process devoid of meaning in a real-world context. It can, however, be approached in a rational manner, as we shall explain later.

ALTERNATIVES AND SUPPLEMENTS TO THE STANDARD ALGORITHM

First it must be stated that the standard algorithm is appropriate only after more understandable approaches have been developed. Some of these will now be discussed.

The Unit-Rate Method

The approach with the most intuitive appeal is undoubtedly the "how much (many) for one," or the unit-rate method. It has intuitive appeal because children have made purchases of one and many things and have had the opportunity to calculate unit prices and other unit rates. In fact, many standard one-step multiplication and division problems from third grade onward can be thought of as determining the unit rate (a division problem) or some multiple of the unit rate (a multiplication problem).

Let's look at two examples:

Example A. Sally paid $0.90 for each computer disk (unit rate). How much did she pay for a dozen? Solution is equal to some multiple of the unit rate = .90 × 12.

Example B. Sally bought a dozen computer disks for $10.80. How much did each disk cost? Solution is the unit rate = 10.80/12 = 0.90.

In the first example the unit rate is given and the student is to find some multiple of it. In the second example the student is asked to find the unit rate, given some multiple of it. Vergnaud (1983) suggests that many one-step multiplication and division problems can be thought of as missing-value ($a/b = c/x$) proportional situations where the unit rate is given or is to be found. Proportional situations where the unit rate is not given are, in fact, the standard missing-value problem types. In the problem that follows, note how a slight modification in the conditions of the problem changes the difficulty level. We now have a two-step problem whose solution encompasses precisely those elements discussed above.

Example C. Sally paid $4.50 for 5 computer disks. How much did she pay for a dozen? We have here a two-part solution: *First* find the unit rate. This requires a division (4.50/5), as in example B. Then find the appropriate multiple of the unit rate [(4.50/5) × 12 or .90 × 12], as in example A.

Thus we see that the format for a standard missing-value problem is very much related to previously learned one-step multiplication and division problems. As the context (setting, rate type, etc.) and the numerical aspects of the problem are adjusted, the problem can be made more (or less) obscure to the student, but the format and the interpretation of proportional events remain unchanged.

This relationship to previously understood material is important pedagogically. Looking back, we see that "new" ideas (proportional situations in the form of missing-value problems) can now be rethought from familiar multiplication and division perspectives. Looking ahead, we see that the unit rate is also interpretable as the slope (m) of a linear function of the form $y = mx$. More will be said about that later.

It must be noted that there are always two unit rates for a given rate pair, each being the reciprocal of the other. One is usually more useful and more easily interpretable than the other. In the previous example, 5 computer disks cost 450 cents. Ratios can be expressed in two ways:

$$\frac{450 \text{ cents}}{5 \text{ disks}} \quad \text{or} \quad \frac{5 \text{ disks}}{450 \text{ cents}}$$

The first is interpreted as 90 cents for 1 disk, the result of the division $450/5 = 90$. The second is interpreted as $0.0111\ldots$ disks per cent, resulting from the division $5/450$. This is a mathematically correct interpretation and useful if we are interested in how many disks could be purchased for $10. (Note that this latter problem is usually solved by dividing $10 by 90 cents.)

This choice causes students great difficulty. In many instances they are unable to interpret reciprocals of standard unit rates, probably because the idea is never formally considered and because they have been taught when dividing to ask how many times the bottom number "goes into" the top number. This leads to further problems when interpreting ratios because "there are no cents in computer disks."

Rates and their reciprocals (like functions and their reciprocals, graphs and their reciprocals, numbers and their reciprocals, etc.) are important mathematical concepts. Considering reciprocals in this relatively concrete context will facilitate later extensions of the concept to algebraic settings. For example, if Sally bought a disks for b dollars, how many disks could she buy for c dollars? Notice that here it is very important that both unit rates be interpretable by students. a/b has a very different meaning from b/a. Which unit rate is appropriate in this problem? Why?

The Factor of Change Method

A second, slightly less functional but very valid method for solving missing-value problems involves a "times as many" mentality. In proportional situations, if one variable is x times another within a given rate pair, this variable should likewise be x times the other in its equivalent rate pair. This is also true between rate pairs. Let's examine the idea using a modified version of our previous example:

Example D. If Sally paid $3.60 for 4 computer disks, how much did she pay for a dozen?

An individual using the factor of change method would reason as follows: "Since I want three times the number of disks, the price should be three times as large, so the answer is 3×3.60." Similarly, one could reason that since 4 is 1/3 of 12, then $3.60 must be 1/3 of the required total cost, that is, 3×3.60. Notice that this argument has underpinnings related to rational numbers and is based on the idea that if I know 1/3 of something, then I can generate that something (the unit or whole) by multiplying by 3, since 3 × 1/3 = 1 (unit).[1]

You may be wondering why we suggested that the factor method was less functional than the unit-rate method. This is so because the ease of use with the factor method is very much related to the numerical aspects of the task. We chose example D because the numbers are very compatible with one another (integral multiples) and lend themselves to a "times as many" approach. What if the problem were like example C? Would we have been as likely to say, "Since 12 is 2 2/5 as large as 5 the price paid should be 2 2/5 as large as $4.50" (i.e., 12/5 × 4.50)? Probably not, although it is still conceivable. We have found that many teachers also have some difficulty with the factor interpretation when the numbers used are not multiples of one another. As would be expected, additional difficulties arise if the relationships are expressed with variables—that is, if Sally paid a cents for b disks, how much did she pay for four times as many ($4b$) disks?

To summarize, the factor method is a useful interpretation of proportionality and one that should be in every child's repertoire. It makes a certain subclass of word problems very easy to solve. Its use is, however, generally confined to those problems where corresponding numerical values in the rate pairs are integral mulitples of one another.

APPLICATIONS TO NUMERICAL COMPARISON PROBLEMS

A second type of proportion-related problem (the first was the standard missing-value problem) involves a comparison of two rates and the determination of which one is less or greater. This problem type is known as a numerical comparison.

Example E.   Mary bought 4 computer disks for $3.60. Joanna bought 5 identical disks for $4.25. Who had the better buy?

Notice that a "how much for one" mentality is a natural for this problem. Two simple calculations (divisions) will generate two unit rates, which then can easily be compared. Joanna had the better buy because she paid $0.85 a disk whereas Mary paid $0.90 a disk. A unit rate approach can be used effectively and is a natural way to deal with all numerical comparison problems. This is yet another reason why it should have a prominent place in the prealgebra curriculum. Our research with junior high school children (Heller, Post, and Behr 1985; Post et al. 1985) has determined that the unit

---

1. There are many situations in mathematics where a portion of the unit is given and the task is to generate the whole. These situations all have antecedents in rational number understandings. It is interesting to note that textbook-based experiences deal with finding parts given the whole but not the other way around. Is it any wonder, then, that so many children have such difficulty with these concepts and later, related ones occurring in an algebraic context, such as $\frac{2}{3}x = 8$?

rate method is far and away the most widely used by children who have not had formal instruction in the standard cross-multiplication algorithm. Ginsburg (1977) suggested that we as mathematics educators should attempt to exploit the mathematics that children already know and attempt to capitalize on and extend the thought processes that occur naturally. The unit rate procedure surely satisfies both these criteria.

THE GRAPHICAL INTERPRETATION OF PROPORTIONALITY

Sets of equivalent rate pairs, ratios,[2] and rational numbers can be represented graphically with the exception that the two axes are exempted. With one exception, $(0,0)$, the vertical axis is uninterpretable, all entries having a zero in the numerator, and the horizontal axis is undefined, since entries on it would imply division by zero. The point $(0,0)$ often has a viable interpretation, that is, zero cost for zero computer disks. Points representing equivalent fractions, ratios, or rate pairs will define a straight line through the origin. The reader will recognize this situation as a linear function of the form $y = mx$. Most proportional situations are typically restricted to the first quadrant because the values generated by "real life" are normally positive. As students become comfortable with interpreting physical situations through graphs in the first quadrant, extensions to more complex and nonproportional phenomena, such as the graphing of equations of the form $y = mx + b$ and nonlinear forms $(xy = k, y = x^2)$, are appropriate.

Graphs can be used to generate equivalent ratios or rate pairs and to identify the unknown in the second rate pair of a missing-value problem. This is so because the linear function, of which the first rate pair is a part, completely defines the relationship between all equivalent rate pairs. A set of all equivalent rate pairs is called an equivalence class. In missing-value problems the second rate pair is always a member of this equivalence class and thus is represented on the same graph (line) as the first rate pair.

The process is illustrated below. Recall exercise C (Sally bought 5 disks for $4.50; how much for a dozen?). This can be represented in the form $5/450 = 12/x$.

To find x graphically in this proportion, first plot the known ordered pair $(5, 450)$. Then connect with the point $(0,0)$. Remember $(0,0)$ means that if there are no disks, then there is no cost, a reasonable proposition. Extend the line (suggested by these two points) through point $(5, 450)$ as in figure 8.1 below. Next locate the desired number of disks (12) on the horizontal axis, proceed vertically until the function line is met, then proceed directly to the

---

2. It is generally agreed that ratios compare like quantities or measures, i.e., 4 dollars/6 dollars = 4/6, whereas rates compare unlike quantities or measures, i.e., 4 miles in 3 hours = 4 miles/3 hours. Ratios can be expressed numerically without labels; rates must retain their labels to be interpretable. Both are manipulated in the same manner.

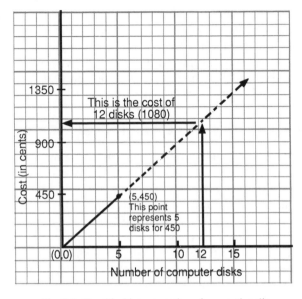

Fig. 8.1. Graphical interpretation of proportionality

verticle axis. The point on the vertical axis (1080) represents the cost of twelve disks. A similar process could be followed for any number of disks. Conversely, the number of disks obtainable for a specific cost could be found by reversing the process.

Notice that the equation of this line is $y = 90x$. This is interpretable as cost = 90 × number of disks. But 90 is also the cost of a single disk—the unit rate. In addition 90 is also the slope of the line and often is referred to as the constant of variation. The graph can be used to identify the unit rate by locating 1 on the horizontal axis and following the procedure outlined above.

The slope of a line denotes the multiplicative nature of the relationship between the variables. Thus it is a very special number. The fact that 90 is also the unit rate (another special number) is no coincidence. The unit rate is always the slope of the line if that slope is expressed with a denominator of 1. You will see in the next section that any rate pair can be changed to the unit rate by simple division.

We already know that the slope (m) of a line of the form $y = mx$ (straight line through the origin) is y/x (solving for m). But y/x is the same division used to define the rate pair and to determine the unit rate (recall example B). It follows, then, that the unit rate and the slope of any line expressed with a denominator of 1 (i.e., a slope of 4 is expressed as $4/1$; a slope of $5/4$ is expressed as $[5/4]/1$) are one and the same.

SOME EMPIRICAL RESULTS

In the spring of 1985, the Rational Number Project (Behr et al. 1983; Post et al. 1985) conducted a survey of over 900 seventh and eighth graders from an outlying suburban district near Minneapolis. Student achievement with missing-value, numerical-comparison, and qualitative-reasoning problems was very low (seventh-grade average was 50% correct; eighth grade, 66%) despite the fact that six of the eight sets of numerical values used were integral multiples of one another. Student responses were examined for strategies used. The unit rate and factor strategies were by far the most successful for the seventh graders and were used in about 30 percent of the problems by the eighth graders. This represented half of all the correct answers, even for students who had been taught the cross-multiplication algorithm.

A SECOND LOOK AT THE STANDARD ALGORITHM

In this section we relate the cross-multiplication algorithm to previous discussions and illustrate how it can be reinterpreted as a shortcut combination approach that employs the same type of thinking used in the unit-rate or factor method.

Example F: Nicole and Colin were running equally fast around a track. It took Nicole 6 minutes to run 4 laps. How long did it take Colin to run 10 laps?

Situation 1: The unit-rate approach. Two appropriate rates are as follows:

$$\text{Nicole's rate} = \frac{6 \text{ minutes}}{4 \text{ laps}}; \quad \frac{x \text{ minutes}}{10 \text{ laps}} = \begin{array}{l} \text{Colin's rate, which is} \\ \text{equivalent to Nicole's} \end{array}$$

Using the standard cross-multiplication algorithm we get

$$\frac{6 \text{ minutes} \times 10 \text{ laps}}{4 \text{ laps}} = x \text{ minutes} \qquad (1)$$

or

$$\frac{6 \text{ minutes}}{4 \text{ laps}} \times 10 \text{ laps}.$$

But what does Nicole's rate of 6 minutes/4 laps signify?

A bit of reflection will suggest that this is nothing more than the unit rate, denoting minutes per lap, which can be rewritten as

$$\frac{1\ 1/2 \text{ minutes}}{1 \text{ lap}}.$$

In fact, any rate pair can be rewritten as a unit rate by performing a simple division. Such a division will always reduce the denominator to 1, which is, by definition, the unit rate. Recall that division was the operation suggested in examples B and C for finding the unit rate. It follows that any rate pair or ratio is the unit rate expressed in a more complicated format, that is, with a denominator not equal to 1.

Now, if Nicole's unit rate is 6 minutes/4 laps (or 1 1/2 minutes per lap), then Colin's time can be found by multiplying this same unit rate (since they were running equally fast) by number of laps that he ran, in this instance ten.

$$\begin{array}{ccc} \text{Nicole's unit rate} & \times & \text{number of laps that} & = & \text{Colin's} \\ \text{(in minutes/lap)} & & \text{Colin ran} & & \text{time.} \end{array}$$

Symbolically we write

$$\frac{6 \text{ minutes}}{4 \text{ laps}} \times 10 \text{ laps} = 15 \text{ minutes.}$$

We now see that the standard cross-multiplication algorithm is an application of the unit-rate approach as discussed earlier.

Situation 2: The factor-of-change method. What if the ratios established were

$$\frac{6 \text{ minutes}}{x \text{ minutes}} = \frac{4 \text{ laps}}{10 \text{ laps}} ?$$

Solving, we again have

$$x \text{ minutes} = \frac{6 \text{ minutes} \times 10 \text{ laps}}{4 \text{ laps}} . \qquad (2)$$

This could be rewritten as in (1), in which case the argument above applies, or it could be rewritten as

$$6 \text{ minutes} \times \frac{10 \text{ laps}}{4 \text{ laps}} .$$

But the ratio $\frac{10 \text{ laps}}{4 \text{ laps}}$ can be rewritten as 10/4, or 2 1/2. (Recall that ratios, unlike rate pairs, are generally expressed numerically without labels.) Colin ran 2 1/2 times as far as Nicole, so his time should be 2 1/2 times as great. The solution to our problem now becomes 6 minutes × 2 1/2 = 15 minutes, obviously the same result.

The type of analysis discussed here can always be used to reinterpret the standard algorithm or cross-multiplication approach as either a unit-rate or factor strategy. This will make it more understandable to students. As a matter of principle, the more that students understand, the more they will view mathematics as an intricate and ever-expanding web of previously learned and interrelated ideas rather than a collection of arbitrary rules that have no apparent relationship or rationale.

A full understanding of the instances of simple (or direct) proportion discussed in this paper is a necessary prerequisite for later involvement with more "awkward" numerical entries; with more remote physical contexts such as density or solubility; with multiple and inverse proportions; and with other situations encountered in algebra, geometry, and the sciences that involve a series of magnitudes (variables) whose interrelationships cannot be represented by the simple linear equation $y = mx$.

CONCLUSION

Algebra is often defined as generalized arithmetic. Students must understand the connections between the abstract equations of algebra and the real world of arithmetic. The concept of using a variable as a placeholder will constitute an important conceptual underpinning for much of what is to follow. Introductory algebra must be based on the notion that variables can be manipulated in a manner that exactly parallels many aspects of the real world. This makes algebra both powerful and abstract. Proportional situations provide an ideal entry into the arena of algebraic representation, since its arithmetic precursors are justifiable through commonsense approaches.

The unit-rate method was suggested as the scaffolding on which other interpretations can be constructed because it is a powerful technique, yet one that capitalizes on children's natural thought patterns and inclinations to use this approach.

This paper has also suggested that approaches to proportionality be addressed from multiple perspectives. This position is firmly grounded in theory (Dienes 1967), and has been used extensively in the research of the Rational Number Project (Behr et al. 1983; Post et al. 1985). Generally, equipping students with a variety of perspectives and solution strategies fosters not only better understanding but also a more confident and flexible approach to problem solving.

REFERENCES

Behr, Merlyn J., Richard Lesh, Thomas R. Post, and Edward Silver. "Rational Number Concepts." In *Acquisition of Mathematics Concepts and Processes,* edited by Richard Lesh and Marsha Landau. New York: Academic Press, 1983.

Dienes, Zoltan P. *Building Up Mathematics.* London: Hutchinson Educational Publishers, 1967.

Gagné, Robert. "Some Issues in the Psychology of Mathematics Instruction." *Journal for Research in Mathematics Education* 14 (January 1983): 7–18.

Ginsburg, Herbert. *Children's Arithmetic: The Learning Process.* New York: D. Van Nostrand Co., 1977.

Goodstein, Madeline P. *Sci Math—Applications in Proportional Problem Solving.* Menlo Park, Calif.: Addison-Wesley Publishing Co., 1983.

Heller, Patricia, Thomas Post, and Merlyn Behr. "The Effect of Rate Type Problem Setting and Rational Number Achievement on Seventh Grade Students Performance on Qualitative and

Numerical Proportional Reasoning Problems." In *Proceedings of the 7th Annual Meeting of the North American Chapter of the International Group for the Psychology of Mathematics Education*, edited by Suzanne Damarin and Marilyn Shelton, pp. 113–22. Columbus, Ohio: PME-NA, 1985.

Karplus, Robert, Steven Pulas, and Elizabeth Stage. "Proportional Reasoning and Early Adolescents." In *Acquisition of Mathematics Concepts and Processes*, edited by Richard Lesh and Marsha Landau. New York: Academic Press, 1983.

Kurtz, Barry, and Robert Karplus. "Intellectual Development beyond Elementary School VII: Teaching for Proportional Reasoning." *School Science and Mathematics* 79 (May-June 1979): 387–98.

Lawson, Anton, and Warren Wollman. "Encouraging the Transition from Concrete to Formal Cognitive Functioning: An Experiment." *Journal of Research in Science Teaching* 13 (1976): 413–30.

Lunzer, E. A. "Problems of Formal Reasoning in Test Situations." *Monographs for the Society of Research in Child Development* 30 (1, Serial No. 100), 1965.

Noelting, Gerald. *The Development of Proportional Reasoning and the Ratio Concept. Part 1—The Differentiation of Stages. Educational Studies in Mathematics II*, pp. 217–53. Boston: Reidel Publishing Co., 1980.

Post, Thomas R., Merlyn Behr, Richard Lesh, and Ipke Wachsmuth. "Selected Results from the Rational Number Project." In *Proceedings of the 9th International Conference for the Psychology of Mathematics Education*, edited by Leen Streefland, pp. 342–51. Utrecht, Holland, 1985.

Vergnaud, Gerard. "Multiple Structures." In *Acquisition of Mathematics Concepts and Processes*, edited by Richard Lesh and Marsha Landau. New York: Academic Press, 1983.

Wollman, Warren, and Robert Karplus. "Intellectual Development beyond Elementary School V: Using Ratio in Differing Tasks." *School Science and Mathmatics* 74 (November 1974): 593–613.

CAN YOUR ALGEBRA CLASS SOLVE THIS?

Problem 8. Find x^2 if x satisfies the equation

$$\sqrt[3]{x + 9} - \sqrt[3]{x - 9} = 3.$$

Solution on page 248

CAN YOUR ALGEBRA CLASS SOLVE THIS?

Problem 9. Given $2^x = 8^{y+1}$ and $9^y = 3^{x-9}$, find the value of $x + y$.

Solution on page 248

Part 3
Equations and Expressions in Algebra

9

Two Different Approaches
Among Algebra Learners

Carolyn Kieran

A LGEBRA is often called "generalized arithmetic." The term implies that
the arithmetic operations are generalized to expressions involving
variables. Thus, expressions like $x + 3$ or $2y + z$ are considered in the same
sense as, for example, $5 + 3$ or $2 \times 7 + 8$, in that x, y, and z represent
numbers. In arithmetic, an addition sign between two numbers always
indicates that the two addends are to the added, as in $5 + 3 = \boxed{?}$. But it does
not always mean this in algebra. In the equation $2x + 5 = 13$, the addition
sign does not mean that the given numerical terms on the left side, 2 and 5,
are to be added. Unless we know the numerical value of $2x$, the addition sign
of this equation signifies that 5 is to be subtracted from 13. In other words,
the operation signs in an equation are not necessarily the operations to be
used in solving the equation. Thus, a major difference between arithmetic
and algebra is this distinction between the operations that are used in the
process of solving algebraic equations and the given operations of these
equations.

How does a student view equations and equation solving in the initial
period of learning algebra? Two different approaches were found to exist
among six thirteen-year-old seventh graders who participated in a three-
month teaching experiment. One of these approaches focused on the *given*

operations; it is called the *arithmetic approach*. The other focused on the *inverses* of the given operations; it is called the *algebraic approach*. The arithmetic approach translated into the use of such solving procedures as trial-and-error substitution. The algebraic approach was characterized by the equation-solving procedure of "transposing terms to the other side."

FIRST INDICATIONS OF TWO DIFFERENT APPROACHES

The students who participated in this work had not yet begun to take algebra in their mathematics classes; they had followed a standard arithmetic program during their years in elementary school. They were all of average mathematical ability. To start with, each student was interviewed. There were three types of questions: questions on (1) different parts of an equation, such as the equal sign and unknown term; (2) equation solving; and (3) equivalence of equations. The students' responses suggested that algebra learners can be divided into two groups as described below. These two groups not only had different preferred methods of solving equations but also different views on the meaning and significance of the various parts of an equation.

Meaning of Letters Used in Equations

The students were asked what the letter means in $5 + a = 12$ and $2 \times c + 15 = 29$. There were two distinct types of answers, and I have classified the students accordingly. Those whose answers referred to the inverse operations necessary to find the value of the letter I have called the "*algebra* group," and those who did not mention inverse operations but rather stated that the letter was a number I have called the "*arithmetic* group." In naming these two groups, I do not wish to imply that the algebra group was the stronger or that their views were more advanced than those of the arithmetic group. I chose the term *algebra group* primarily because of the focus of these students on inverse operations.

For the algebra group, the letter seemed to have meaning only when its value was found. A typical answer to the question, "What does the letter mean in $5 + a = 12$?" was "An answer—12 minus 5 is 7." For them, an equation had to be reformulated for the letter to have some meaning, as in, for example, $5 + a = 12$ being transformed to $a = 12 - 5$, or $2 \times c + 15 = 29$ to $c = (29 - 15)/2$.

The arithmetic group, however, seemed to view the letter as standing for some unknown number. In that sense, the letter was part of the numerical relationship of the equation. These students expressed the unknown in the equation using the given operations. For example, a typical answer of this group to the question, "What does the letter mean in $5 + a = 12$?" was "It means a whole number that's going to be added to 5 and it's going to equal 12."

These two different perceptions of the letter in an equation were subsequently made more explicit by asking the meaning of the letter in the expression $a + 3$. The arithmetic group said that the letter was "some number" or "any number." However, the algebra group could not assign any meaning to the letter. This can be explained by the fact that when a letter is viewed in terms of inverse operations, this interpretation cannot be used with algebraic expressions such as $a + 3$. Since expressions lack an equal sign as well as a number following it, the "algebra" student cannot apply any inverse operations, and hence the letter has no meaning in this context.

Equation Solving

Though the students had not yet begun to take algebra, they were nevertheless able to solve some simple equations. The procedures that they preferred to use corresponded with the way they viewed the letter in equations. The arithmetic group used the given operations to solve equations, substituting different numbers for the letter until they found one that balanced the left side and right side of the equation. The algebra group used the inverses of the given operations and solved by transposing terms to the other side. They even transposed very simple equations such as $4x = 20$, for which they said that "x is 20 divided by 4." Their solving approach was basically to take the number on the right side and to use the inverse of each operation as they came to it, going in a right-to-left direction.

Equivalence of Equations

The preteaching interview also included questions on the equivalence of equations. One such question was "What happens if I add 10 to the left side of $x + 469 = 1351$?" The arithmetic group suggested adding 10 to the other side to balance the equation, that is, $x + 469 + 10 = 1351 + 10$. Students in the algebra group suggested subtracting 10 from the other side, that is, $x + 469 + 10 = 1351 - 10$. They believed that x would "work out the same" in both the given and transformed equations.

Additional Difficulties of Algebra Group before Instruction

An error committed by some students in the algebra group was overgeneralizing the right-to-left transposing procedure in order to handle equations with two unknowns on the left side. They began with the term on the right and transposed from right to left. For example, when they were asked how to find the value of a in $3a + 3 + 4a = 24$, they said "24 divided by 4, minus 3, minus, uhm, no, divided by 3." Obviously, this overgeneralization had led to a dilemma. After dividing by 4 and then subtracting 3, they did not know whether they should subtract or divide by the remaining 3. That is, there were two remaining operations of which they had to take the inverse, but only one numerical term.

Overgeneralization of the right-to-left solving strategy also occurred with multioperation equations containing an unknown on each side of the equal sign, as in $2 \times c + 5 = 1 \times c + 8$. Again there was the problem of not knowing which operation to "inverse," since there were more operations than numbers, that is, whether to do 8 minus 1 or 8 divided by 1. They usually overcame such dilemmas by ignoring the addition operation and by taking the inverse of the multiplication.

It is important to note that the overgeneralized procedure of beginning with the number on the extreme right and then moving from right to left, taking the inverses of the operations as one moves along the sequence, clearly assigns very little significance to the role of the equal sign within the equation-solving process. The existence of this approach raises some serious questions with respect to the degree of emphasis we should be placing in elementary school on the use of transposing as a procedure for solving open sentences.

THE TEACHING EXPERIMENT

After the preliminary interview, the students participated in approximately ten one-on-one teaching/learning sessions. These sessions were designed to enlarge the students' notions of *equation, unknown,* and the *equal sign* and to teach them an equation-solving procedure that made explicit the left-right balance of an equation. The instruction was based on an approach described in a past issue of the *Mathematics Teacher* (Herscovics and Kieran 1980). Though these students are from another project, the teaching approach started off the same way. You are invited to refer to that article for details.

In brief, the students began by generating "arithmetic identities" and equations. The approach emphasized the importance of the equal sign as a symbol of the equality relationship between the left and right sides of an equation (Kieran 1981). It also led the algebra group, who had not used trial-and-error substitution in the preliminary interview, to begin using this method to solve certain equations.

After reaching an understanding for the procedure of doing the same operation on both sides of an arithmetic identity, the students spent time attempting to make the transition from carrying out these operations on arithmetic identities to carrying them out on equations. The solving procedure of performing the same operation on both sides of an equation was the focus of the teaching experiment.

It should be noted that the way in which inverse operations are used in this procedure is quite different from the way they are used in the transposing procedure. Though transposing is often considered to be a shortened version of performing the same operation on both sides, observations of the students in this project suggest that these two procedures may be seen as quite dissimilar by students. The method of performing on both sides of the equation an operation that is the inverse of one of the given operations makes explicit the left-right balance of the equation. Furthermore, the justification for performing the same operation on both sides is precisely to keep the equation in balance and to maintain the solution unchanged throughout the equation-solving process. Moreover, this procedure also involves simplifying both left and right sides of the equation rather than just one side, which occurs when one transposes terms to the other side. This emphasis on the left-right balance of an equation is absent in the transposing procedure.

Though the students of the algebra group seemed to acquire the notion of a letter as number and reinforced this notion by using the substitution procedure for certain equations throughout the teaching experiment, they resisted using the procedure of performing the same operation on both sides. They seemed unable to make sense of it and preferred instead to extend their transposing procedure and to use it whenever possible. By the end of the teaching experiment, only half the students were regularly using the procedure of performing the same operation on both sides. These were the students of the arithmetic group.

PEDAGOGICAL IMPLICATIONS

The way the students' solving methods evolved throughout the project suggests that three sublevels may be involved in learning the procedure of performing the same operation on both sides. The first is reconstructing one's view of the letter in an equation to encompass the notion of letter as number within the given sequence of operations. The second involves translating this conceptualization of a letter into the related equation-solving procedure of trial-and-error substitution. The third requires replacing the procedure of substitution by the procedure of performing the same operation on both sides.

Since this work was carried out with students who were just beginning high school, there are some clear pedagogical implications for instruction in elementary school mathematics. The problem of solving open sentences— for example, $\square + 3 = 7$, or $n + 3 = 7$—is often handled in elementary school by referring to the corresponding inverse equality $7 - 3 = \boxed{?}$. This approach may be leading to a view of an algebraic letter as the result of a set of inverse operations—with the letter having no existence of its own within the given equation. These observations suggest that the current approach used in most elementary school mathematics programs may be inadequate. Elementary experience with placeholders ought to include a conceptual approach to equations like the one adopted in this study, one where the emphasis is not on giving meaning to the letter by transposing the given equation but rather on giving meaning to the letter as a number within the given sequence of operations of the equation. Consequently, finding the value of the place-holder in an open sentence is accomplished by the substitution method. This kind of experience at the elementary school level might help to enhance the beginning algebra student's ability to make sense not only of the solving procedures taught in high school but also of equations themselves.

Another current practice in the teaching of equation solving that may be leading to difficulties in algebra is the use of the "Think of a number" game, which goes like this: "I am thinking of a number. If I multiply it by 4 and add 2, the result is 50. What is my number?" Often students are advised to undo the given operations in reverse order to find the original number. However, such an approach may simply be reinforcing a notion of letter such as that found among the algebra group and may in fact be encouraging the right-to-left transposing errors we saw earlier.

In conclusion, this experience has uncovered among beginning algebra students two different perspectives on equations and equation solving. Those who belonged to the algebra group at the beginning of the project continued to prefer to use the transposing solving method at the end. Those who belonged to the arithmetic group at the beginning of the project were the ones who were able to make sense of the taught procedure of performing the same operation on both sides and who actually used this method by the end of the project. These findings suggest that the construction of meaning for this solving procedure may be a learning process that takes time. As mentioned earlier, consideration ought to be given to the idea of starting this process in elementary school. In the meantime, high school teachers might take into account the likelihood that some of their beginning algebra students may prefer the procedures used by the "algebra group" and that others may prefer those of the "arithmetic group." Instruction that provides for both of these preferences will probably be more successful than instruction that is geared to just one of them.

REFERENCES

Herscovics, Nicolas, and Carolyn Kieran. "Constructing Meaning for the Concept of Equation." *Mathematics Teacher* 73 (November 1980): 572–80.

Kieran, Carolyn. "Concepts Associated with the Equality Symbol." *Educational Studies in Mathematics* 12 (August 1981): 317–26.

CAN YOUR ALGEBRA CLASS SOLVE THIS?

Problem 10. What are the values of k for which the equation $2x^2 - kx + x + 8 = 0$ will have real and equal roots?

Solution on page 248

10

An Integration of Equation-solving Methods into a Developmental Learning Sequence

John E. Bernard
Martin P. Cohen

EQUATION solving was the main theme of the work *Hisab al-jabr w'al-muqabalah*, by Mohammed ibn-Musa al-Khowarizmi (A.D. 825), which marked the official beginning of algebra. Since equations play such a central role in mathematics and many of its applications, it is little wonder that learning to solve equations is still an essential element in the study of algebra. In this article we discuss the development of the basic equation-solving skills typically taught in first-year algebra. Our chief concern is with how teachers, through an understanding of students' thought processes and perceptions of the equation-solving task, might present a sequence of appropriate learning experiences designed to help students formulate and perfect their knowledge and skill.

The discussion is limited to very basic linear equations in one unknown with rational coefficients and with each member constituting a well-formed algebraic expression. Figure 10.1 shows an assortment of the intended type of equations. The guiding questions for our discussion are these: (1) What is a root of an equation? and (2) How does one go about finding an unknown root?

WHAT IS A ROOT OF AN EQUATION?

The question concerning the nature of a root is answered reasonably well in modern texts. Usually, the variable of the equation is associated with an appropriate domain. Next, a special subset of the domain is identified consisting of all those values and only those values that make the equation

---

1. $5 - n = 2$

2. $\dfrac{7 - x}{2} = 2$

3. $17 = \dfrac{y}{4} + 19$

4. $25 - 3w = 2w$

5. $3a + 6 = 2(9 - a)$

6. $25 - (15 - n) = 2$

7. $14x - x = 11(70 - 2x)$

8. $12 - \dfrac{x + 5}{10} = \dfrac{3x}{5} + 15$

Fig. 10.1. Linear equations in one unknown typically found in first-year algebra

true on substitution and subsequent evaluation. This subset is known as the truth, or solution, set of the equation. Each element of the solution set is called a *solution*, or *root*, of the equation. Hence, a root is a testable value that makes the equation true, and solving is the process of finding such values.

This definition of *root* indicates the basic knowledge a student needs to begin a meaningful study of solving equations. It entails a knowledge of numbers and variables, facility with substitution and computation, and an ability to judge whether or not the left and right members of an equation are equal.

To introduce learners having these prerequisites to the solving of equations, we suggest using an interactive computer program like that shown in figure 10.2. With computation automated, students can concentrate on identifying roots and nonroots. The criterion by which the distinc-

```
100 INPUT "PLEASE INPUT THE NUMBER TO BE TESTED;" X
120 LEFTSIDE = 3 * X + 6
140 RIGHTSIDE = 2 * (9 - X)
160 IF LEFTSIDE < > RIGHTSIDE THEN 240
180 PRINT X; " IS A SOLUTION."
200 PRINT "PRESS ANY KEY TO CONTINUE"
220 GET Z$ : GO TO 300
240 PRINT X; " IS NOT A SOLUTION."
260 PRINT "PRESS ANY KEY TO CONTINUE"
280 GET Z$
300 PRINT "PRESS E TO END, ANY OTHER KEY TO TEST A NEW
    VALUE."
320 GET Z$: IF Z$ < > "E" THEN HOME: GO TO 100
340 HOME: VTAB 12: PRINT "THANKS, HOPE TO SEE YOU AGAIN"
360 END
```

Fig. 10.2. A BASIC interactive program for testing possible solutions for $3x + 6 = 2(9 - x)$. Change lines 120 and 140 appropriately to adjust for other equations.

tion is made (fig. 10.2, lines 160, 180, and 240) should be especially singled out and stressed in helping students formulate their initial understanding of roots and the solution process.

The program also introduces *checking*. The features of (1) accepting a possible root, (2) evaluating each member of the equation separately, and (3) determining whether the results are equal or not provide an enactment of checking. Thus, checking is a natural part of solving equations. This should be pointed out especially to those students who have a tendency to end their equation-solving work prematurely with statements such as $x = 2$. Checking should bring solvers the satisfaction of closure that does not depend on the teacher.

HOW DOES ONE GO ABOUT FINDING AN UNKNOWN ROOT?

There are many ways to find unknown roots—many methods for solving equations. Those discussed here, including the prominent *equivalent equations* method taught in the traditional first-year algebra course, are not presented in isolation. Instead, we show how to use them to construct a *developmental learning sequence*, where each subsequent method of solution evolves from its predecessor. The construction is done in such a way that students' existing capabilities and understanding of the task lead to the development of each new method. In addition, we identify the knowledge and understanding needed for the meaningful learning of each new method. The methods in the learning sequence are (1) *generate and evaluate*, (2) *cover up*, (3) *undoing*, and (4) *equivalent equations*.

The Generate-and-Evaluate Method

Students armed with an understanding of a root of an equation, as depicted by the computer program in figure 10.2, and having a problem-solving outlook can be led to develop the generate-and-evaluate method. Given a basic equation, our solvers would be cued to think of their concept of numbers and to generate different values for testing. Near the beginning of their efforts, this mental generating might appear to be random, but it is not likely to remain so for long. Most early adolescent solvers have some understanding of the linear order of "their" number system and, also, a notion that "more numbers always exist between any two given numbers." Thus, rarely do they continue to operate very long in a trial-and-error fashion.

To illustrate the kind of changes that might take place, consider the case of solving $29 - 4x = 3$. After generating and testing sample values of, say, 5 and then 1, thereby getting 9 versus 3 and 25 versus 3, respectively, some solvers will begin to feel that they are better off using 5 instead of 1. Intuitively they might first set 1 as a bound; second, identify the direction to move for generating new values; and third, consider that the root might be in the neighborhood of 5. Those who might test 4 next, getting 13 versus 3, would realize intuitively that the root is larger than 5. What gradually evolves is an evaluation scheme that feeds back on the generation process in such a way that newly generated values exhibit progress. This full-cycle, recursive approach is the *generate-and-evaluate method*.

The Cover-Up Method

Seldom would teachers be satisfied with leaving students at a trial-and-error or generate-and-evaluate stage of development for very long. We present the cover-up method as a natural next level of development and show how it fits into the bigger scheme of things.

If only equations like those in figure 10.3 are under consideration, generate-and-evaluate equation solvers might easily be guided through a development of the cover-up method.

1. $6950 + x = 6990$	2. $x - 4 = 6$	3. $95 + x = 99$
4. $3x = 18$	5. $5 + x = 9$	6. $15 + (10 - x) = 22$
7. $10 - x = 7$	8. $695 + x = 99$	9. $2(5 + x) = 18$
10. $15 - (10 - x) = 8$	11. $\frac{x}{9} = 9$	12. $\frac{3(x - 4)}{5} = 6$

Fig. 10.3

Present the student with the first five equations one at a time. A switch in method (away from generate and evaluate) will probably occur at least by the time the fifth one is reached. We might well hear, "Oh, that's easy! Five plus four is nine" or "We don't have to solve it; we just know the answer."

Then present the entire selection of equations in figure 10.3 and ask students if they can find other equations like them. This would help to establish and extend the range for their concept of this type of equation (qualified in terms of the solution process). Here, we would expect at least items 2, 4, 7, and 11 to be so identified. If so, we suggest saying to the student, "So, for some equations you can find the unknown just by thinking about what the equation says, what it asks for."

There are several intentional usages here. *Some* is meant to open up the possibility that there may even be more equations of this type than those just identified. This leads, therefore, to a further categorizing of equations according to the methods by which they can be solved. "What it asks for" is a lead-in to the standard phraseology that is used in coaching students through the cover-up method and a link to the meaning of the equation-solving task.

We might present students with

$$14 - x = 8$$

and, while covering up the x, say, "Fourteen minus what number gives you

8?" After the expected answer of 6, we would ask for agreement that this equation could be solved the same way. We then suggest physically extending the equation by writing $16 - ($ $)$ to produce

$$16 - (14 - x) = 8.$$

Ask students to check if we can use our new method for this equation. To explicitly bring about their agreement, convince them that if x varies, then so does $(14 - x)$. Thus, the quantity $(14 - x)$ might be treated as an unknown; that is, just like an x. Now cover the $(14 - x)$ and say, "Sixteen minus what number gives you 8?" After the expected answer, ask, "So what is the value of this quantity that is covered?" (Expected answer, 8) "How can we write that down?" Write—

$$14 - x = 8$$

as a second step. Finishing up, the work looks like this:

$$16 - (14 - x) = 8$$
$$14 - x = 8$$
$$x = 6$$

Does 6 really work? Yes, the check shows that it does:

$$16 - (14 - 6)$$
$$16 - 8$$
$$8$$

The point, of course, is to break the stereotype that an unknown is a single letter and to arrive at a broader conceptualization that an expression containing a variable might also be treated as an unknown. We have also broken the impression that the cover-up approach may be applied only once by showing it being applied iteratively as students gradually make headway from one relatively complicated state to one of lesser complexity and finally identify the unknown value. Recall the educational importance of learning the concept of iteration. This is an appropriate time to teach students the name, *cover-up method*, so they will think of the processes above when they hear it.

As a final touch, guide students to recognize the domain of applicability for the new method, at least to the extent of referring once again to the equations in figure 10.3. Of course, all the equations in the figure might be solved by this method, but what about $3x + 2(5 - x) = 7$, or $10 - 4x = 7 - 2x$? Probably not. The method has some limitations and other methods should still be sought.

The Undoing Method

The *undoing* method is closely related to the *cover-up* method, but it is neither mathematically nor psychologically equivalent to it. The undoing method builds on the notions of operational inverses and the reversibility of a process involving one or more invertible steps. It can be identified with functions and their inverses. The undoing method is related to finding inverses of composite functions. (These rather sophisticated ideas are mentioned only to guide the discussion; they are not intended to be taught explicitly to students in first-year algebra.)

Most, but not all, equations that can be solved by the cover-up method can also be solved by undoing. Consider subtractions of an unknown amount and divisions by an unknown amount, as in these examples:

$$14 - x = 8$$
$$\frac{108}{x} = 9$$

In the first example, the "adding back" of x is not typically done in pedagogical demonstrations of the undoing method. However, any equation that could be solved by undoing could be approached also with the cover-up method; thus, we shall treat undoing as a way to handle some special equations that are resolvable by the cover-up method.

The undoing method might be developed from the cover-up method as the complexity of the equation increases. Consider

$$\frac{7(2x - 3) - 5}{10} = 3.$$

Many students would be interested in skirting the drudgery of writing all the intermediate steps like

$$7(2x - 3) - 5 = 30.$$

Raise this issue with students to open a discussion designed to lead to developing the undoing method.

Guide students to think about what is operationally happening to x. As the steps unfold, teacher and student should keep a record of the activity. The order of operations for evaluating the left member for given x values dictates the sequence in figure 10.4; the resulting 3, of course, is from the statement of equality.

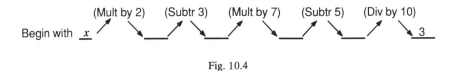

Fig. 10.4

The usual procedure can now be generated with leading questions of how to get back to the starting point, that is, to x. Question: "How did we get to the final 3?" (Expected answer: Divided by 10.) "Is there any way to undo that to get back to what we had before dividing? If so, what?" (Expected answer: Yes, multiply by 10.) "So you had 30 before that . . ." Completing this pattern gives the sequence shown in figure 10.5.

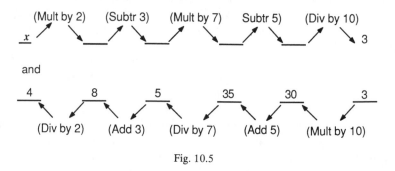

Fig. 10.5

Hence, x, the root, is equal to 4. Spaces rather than algebraic expressions like $2x$, $2x - 3$, and so on, are written so that the check can be made using the originally written sequence. It might be sound practice to demonstrate occasionally the construction of the left member through the use of such intermediate expressions, but there should not be a pedantic insistence that students do this every time.

Even with its limitations, the undoing method is important and should be in the curriculum. It provides an instance of integration within mathematics as it links the solving of equations to previous learning, including the pairing of inverse operations. It thereby reinforces this very important piece of knowledge—important in fostering reversibility, analysis, and problem solving. It demonstrates to learners how their existing knowledge may be used to acquire new capabilities and concurrently provides prerequisites that can be used in learning the equivalent-equations method.

THE CONCEPTUALIZATION OF EQUIVALENT EQUATIONS

The developments discussed so far do not, in our opinion, provide an adequate base for learning the equivalent-equations method. Undoing corresponds to the type of equation that can be characterized as an action sequence of invertible steps leading from a starting point to an ultimate goal. The cover-up method leads into a succession of nested arithmetic facts. And the generate-and-evaluate method depicts an equation as a sentence with variable truth. These conceptualizations do not imply the equivalent-equations method for solving equations.

We know, however, that developing some understanding of solution-set–preserving operations is in order. The *balance operations*, which became prominent elements of the school mathematics curriculum during the "new math" era of the 1960s, certainly qualify as such, but these should not be used to preclude other possibilities. Transpositions, adding zero to or within one or both members, and multiplying any one or more legitimate subexpressions of one or both members of an equation by 1 also qualify. Thus, we propose a few sets of intermediate activities before directly engaging students in the act of solving equations with balance operations. The earlier activities introduce the balance operations. Later activities then promote their use in solution-set–preserving relationships between equations.

Physical devices such as teeter-totters and balance beams have often been used to inspire or justify the legitimacy of "adding the same thing to both sides" and the other balance operations. We have used and recommend such devices (see Goodman et al. [1984]), but here, we shall introduce the symbol T___T (it looks something like a balance beam) to serve our purpose.

T___T is used between two algebraic or numerical expressions to indicate that they are to be compared. The symbols $<$, $=$, and $>$ may then be written in the center space to identify the correct relationship. The symbol T___T may be used in comparison exercises such as those that follow.

Item 1: Determine the larger, or heavier, side in each of the following comparisons. Complete the sentence correctly by writing $<$, $=$, or $>$ in the center space of the T___T.

a) $65 - 9$ T___T $65 - 10$ **b)** $24/8$ T___T $24/6$

Expected answers:

a) $>$ **b)** $<$

Item 2: Correctly complete each comparative sentence. Use $<$, $=$, or $>$.

a) $(27 + 2)$ T___T $(30 - 1)$ **b)** $3(4) + 2 - 2$ T___T $3(4) + 2$

Expected answers:

a) $=$ **b)** $<$

The simplicity of the computation in these items allows students to concentrate on learning the new symbolism. We now hope to "inspire" the balance operations per se through a regulated use of such comparisons and the cover-up approach to completion items. A special sequence follows:

Item 3: Replace the question marks with either numbers or operations to complete the following comparisons so that equality results.

a) $16 + 2$ T___T $16 + \underline{\ ?\ }$ **b)** $9 - 5$ T___T $\underline{\ ?\ } - 5$

c) $100 - 5$ T___T $100 \underline{\ ?\ } 5$ **d)** $3(9)$ T___T $3(5 + \underline{\ ?\ })$

$$\overset{+\ \text{or}\ -}{}$$
e) $42 - 7$ T___T $3(6 + 8) \underline{\ ?\ }\ \underline{\ ?\ }$

f) $\frac{1}{2}(28 + 1)$ T___T $\underline{\ ?\ }(30 - 1)$ **g)** $\dfrac{- 13(6)}{- 13}$ T___T $\dfrac{13(- 6)}{?}$

h) $3(72 - 9) + (8 - 2)$ T _____ T $200 - 20 + \overbrace{9}^{+\ or\ -}$? ?

Hint: $3(72 - 9) = (200 - 20 + 9)$

i) For $3x + 2 = 17$,

$(3x + 2) - 2$ T _____ T $\overbrace{17}^{+\ or\ -}$? ?

Parts (a), (b), and (c) of item 3 merely set the stage in terms of format. Parts (d) through (h) tend to call for close analysis and some reflection on determining the equivalence of subexpressions in the comparisons. These, of course, lead to part (i), which contains a variable and for which the given $3x + 2 = 17$ is significant. Parts (a) and (h) correspond to the addition property; (b), (c), (e), and (i) to the subtraction property; (d) and (f) to the multiplication property; and (g) to the division property of equality. These ought to be discussed with learners in terms of "doing the same thing to both sides."

At this stage you can introduce a few exercises such as those that follow to give the students practice in using the balance operations.

Item 4: Use the addition property of equality to add the given expression to both sides of the equation. Simplify each member separately for the final answers.

a) $8x - 5 = 27$ Expression to be added: 5

b) $7 - 2x = 14 + x$ Expression to be added: $2x$

c) $3(x + 1) - 8 = x + 1$ Expression to be added: $-(x + 1)$

Relatively fluid performance with balance operations should facilitate the discussion of their solution-set–preserving qualities. Our concern is as much, or more, with generating the concept of the solution-set–preserving operation as it is with verifying, or convincing students, that the balance operations have this quality. For the most part, the only additional ideas needed are those identified earlier in connection with understanding roots, that is, students will need to be able to check selected values to determine whether they are roots. We will actually give the roots so that solving the equations will not distract attention from the main objective. Now all that is needed is to set the stage so that the balance operations stand as relational links between equivalent equations. An easy way to do this is through multiple-choice items like the following.

Item 5: The solution set for the equation $3x + 7 = 22$ is $\{5\}$. Encircle the equations that are *not* equivalent to $3x + 7 = 22$. *Hint:* Use the balance operations to eliminate alternatives whenever you can.

a) $(3x + 7) - 7 = 22 - 7$ **b)** $\dfrac{3x + 7}{3} = \dfrac{22}{3}$

c) $3x + 9 = 24$ **d)** $\frac{1}{3} \cdot 3x + 7 = \frac{1}{3} \cdot 22$ **e)** $5 = x$

Item 6: Encircle the equation that is *not* equivalent to $2 - 8x = -(4 + 6x)$, which has $\{3\}$ as its solution set.

a) $8x - 2 = 4 + 6x$ **b)** $-2x = -6$
c) $2 - 2x = -4$ **d)** $2 - 8x = -4 - 6x$
e) $6x - 2 = 4 + 8x$

To reinforce the relational qualities of the balance operations and project the notion of equivalent equations, we suggest the exercise shown in item 7. This exercise represents another step forward, for as one can see from the solution, the equivalent equations in a particular group can be ordered to form a step-by-step pattern like those of the equivalent-equations method. Teachers, guided by their understanding of equivalence classes and properties of equivalence relations like the transitive property, ought to discuss the interrelationships between equations "connected" by means of the balance operations.

Item 7: Here [fig. 10.6], you are given 20 equations; some are equivalent to each other (i.e., have the same solution set), and others are not. Use loops to make separate groups of equivalent equations so that all equations in the

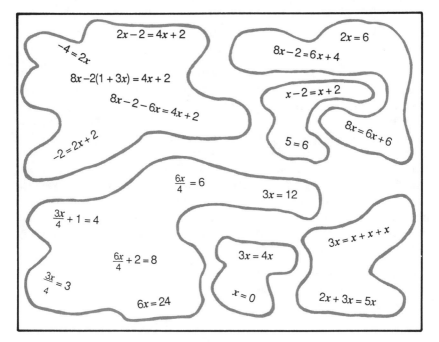

Fig. 10.6

same group are equivalent to one another. [The "answer" is shown with dotted loops.]

We recall the *undoing* method presented earlier to serve as a basis here. This point needs to be made clear that the equations initially presented will be easy and manageable with undoing alone. However, equations not re-solvable by undoing, such as $14 - x = 8$ or more complicated equations with multiple occurrences of the unknown and with unknowns in both members (see fig. 10.1), *do exist!* The learning objective is to develop a new, more general and systematic method for solving equations, not simply to solve some particular equation.

Some reactivation of prerequisite skills might be helpful, such as facility with associating, commuting, combining like terms, removing parentheses, simplifying, and working with signs.

When our students are armed with a knowledge of the balance operations, skill in executing them, and an understanding of how to apply such operations to equations and still maintain their solution sets, we are in a position to present strategies for using the operations to solve equations. Thus, we think we have provided necessary constructive links from the basic conceptualization of an equation and its root(s) to the method for solving equations usually taught in a first course in algebra, which is presented next. Since the method is so common, however, we shall omit some of the details and present the highlights.

The Equivalent-Equations Method

First we pose the following task.

Item 8: Show how to use balance operations to solve

$$2x - 3 = 11.$$

Hint: Think in terms of undoing.

The usual "undoing" analysis, beginning with x, identifies the two operations—"multiply by 2" and "subtract 3"—which are done in this order. Thus, with undoing, "add 3" must be done first. "How can this be done using balance operations?" (Expected answer: "Add 3 to each side through the addition property of equality.") The "subtract 3" operation may then be crossed off the list as the equation is simplified to $2x = 14$. The latter can be "handled" by dividing both sides by 2. Depending on students' ability, however, and to promote the development of a more sophisticated notion of number, the option of multiplying both sides by $1/2$ might be considered. Either way, $\{7\}$ is the solution set for the equation above.

A notable obstacle, however, does exist: The only genuine obstacle to the undoing or cover-up approaches that cannot be resolved by simplifying individual members of linear equations or through some clever manipulations of signs or divisors is that of having unknowns in both members of the equation. Students might be convinced of this if simply given their lead to solve an equation such as $3x - 25 = 2x$ with an undoing strategy. They might discover this obstacle after producing and thinking about either of these equations, neither of which explicitly identifies the unknown root:

$$x = \frac{1}{3}(2x + 25) \quad \text{or} \quad x = \frac{1}{2}(3x - 25)$$

Students could be reminded of prior work with combining like terms in such examples as $3x - 2x = 25$, which removes the "multiple occurrences" obstacle to undoing. But it must be noted that this can be used only when the occurrences are associated with each other on the same side of the equation. So a new subgoal, that of "getting the variables to one side" can be identified. A useful variation, of course, is "getting rid of the variables from one of the sides." The latter relates fairly easily to the disappearance of things like the 3 in $2x - 3 = 11$ as 3 is added to both sides to give $2x = 14$. Aware of this and prompted to think in terms of using the balance operations, the "clever" student will suggest adding $-2x$ to both sides in $3x - 25 = 2x$. One might continue as follows:

$$(3x + -25) + -2x = 2x + -2x$$
$$-2x + (3x + -25) = (-2x + 2x)$$
$$(-2x + 3x) + -25 = 0$$
$$(-2 + 3)x + -25 = 0$$
$$x + -25 = 0$$
$$x + (-25 + 25) = (0 + 25)$$
$$x = 25$$

Thus, since $3(25) - 25 = 50 = 2(25)$, $x \in \{25\}$.

The equivalent-equations method thus presented entails a number of things. Balance operations are used as tools for manipulating the equation with the confidence that the roots are never lost whether or not progress is being made. (*Note:* This is of particular interest if it is further realized that changes do occur in the statements; for example, for $x = 7$, $2x = 14$ corresponds to $14 = 14$, whereas the statement $2x - 3 = 11$ corresponds to $11 = 11$!) And, a sequence of *higher level* processes such as "getting all the variables to one side," "collecting like terms" (to arrive at a single occurrence of the unknown), and "isolating the unknown" is used to guide the *progress* from structurally more complex states of the relationship to later ones of ultimate simplicity.

A comparison between the work of a novice and that of an experienced student in figures 10.7 and 10.8 will help reveal the existence and nature of these higher-level processes. Although figure 10.7 might appear elaborate—and mathematically and psychologically restrictive by today's stand-

Statements	Reasons
1. $3(x - 4) = 2x + 2$	1. Given
2. $3(x + -4) = 2x + 2$	2. Definition of subtraction
3. $3x + -12 = 2x + 2$	3. Distributive property for multiplication over addition, multiplication fact
4. $(3x + -12) + 12 = (2x + 2) + 12$	4. Addition property of equality, adding 12 to both sides
5. $3x + (-12 + 12) = 2x + (2 + 12)$	5. Associative property for addition
6. $3x + 0 = 2x + (2 + 12)$	6. Property for the addition of opposites
7. $3x = 2x + 14$	7. Property of the additive identity, addition fact
8. $3x = 14 + 2x$	8. Commutative property for addition
9. $3x + -2x = (14 + 2x) + -2x$	9. Addition property of equality, adding -2 to both sides
10. $3x + -2x = 14 + (2x + -2x)$	10. Associative property for addition
11. $(3 + -2)x = 14 + (2 + -2)x$	11. Distributive property for multiplication over addition
12. $1x = 14 + 0x$	12. Addition fact, property for the addition of opposites
13. $x = 14 + 0$	13. Property of the multiplicative identity, property for multiplication with zero
14. $x = 14$	14. Property of the additive identity

Check: $3(14 - 4)$ vs. $2(14) + 2$

$3(10)$ vs. $28 + 2$

30 = 30

Therefore, $x \in \{14\}$.

Fig. 10.7. Step-by-step logical (axiomatic) solution for the novice for $3(x - 4) = 2x + 2$

ards—it is representative of pedagogical methods introduced in the "new math" movement of the 1960s. This probably went by the wayside as common sense prevailed. Experienced students usually solve such

Statements	Reasons	Higher-Level Processes or Strategies
1. $3(x - 4) = 2x + 2$	1. Given	"Removing"
2.	2.	parentheses
3. $3x - 12 = 2x + 2$	3. (Close to Novice)	
4.	4.	
5.	5.	"Getting rid" of
6.	6.	the 12
7. $3x = 2x + 14$	7. (Might skip)	
8.	8.	
9.	9.	
10.	10.	Getting the variables
11.	11.	together on
12.	12.	one side
13.	13.	
14. $x = 14$	14. (Might feel "finished")	
Check:		
Therefore, $x \in \{14\}$. (May feel this is trivial or redundant)		

Fig. 10.8. Step-by-step logical solution for the "pro" for $3(x - 4) = 2x + 2$

equations, roughly speaking, in a mere four steps (see fig. 10.8). The perplexing concern for mathematics educators is how to effect the passage sensibly and smoothly from the novice's atomistic level to the more powerful and efficient performance of the pro. Here, we point out that it does happen, it has been happening for generations, and it should happen. We hope that sensitive, sensible teaching and support research will, in due course, resolve this dilemma. (See Bernard and Bright [1984] for a discussion of the transition.)

Note, too, that the higher-level processes (see fig. 10.8) exhibited by the pros are often qualitatively different from the corresponding sequence of logical steps being replaced. In other words, even though faster execution and mental computation explain a portion of the differences, they do not give a totally accurate account. The reliable regularity of patterns in the symbolism provides opportunities during repeated experience for learners to invent effective and efficient task-specific operations or processes that have distinct psychological existence of their own.

Another concern is that in the recent history of mathematics teaching this skill has not been presented with any sense of the problem-solving tradition from which it possibly evolved (a concern for the implementation of NCTM's [1980] *Agenda for Action*). Instead of teaching aimed at students' memorizing a sequence of higher-level processes that they might then rotely execute, it would be better to enact a problem-solving approach that identifies obstacles and subgoals. This would help students develop processes for removing obstacles and attaining subgoals, thus fostering a means for monitoring and evaluating progress and keeping the task and its successful completion at the forefront.

CONCLUDING REMARKS

Mathematics should be taught for relational understanding. This includes equation solving and, in particular, the equivalent-equations method for solving equations. We acknowledge and respect learners' needs to make sense of what they are learning and suggest that they be given the opportunity and assistance needed to construct their new knowledge and to develop their new capabilities as meaningful extensions of what they already know and can do. We have thus presented a synthesis of equation-solving concepts and methods into an organized whole that can be used for the development of equation-solving skills from this viewpoint. It is our hope that this effort will join with others that keep us moving forward in improving mathematics education.

REFERENCES

Bernard, John E., and George W. Bright. "Student Performance in Solving Linear Equations." *International Journal of Mathematical Education in Science and Technology* 15 (1984): 399–421.

Goodman, Terry A., John E. Bernard, Martin P. Cohen, and Joanne E. Meldon. *A Guidebook for Teaching Algebra.* Newton, Mass.: Allyn & Bacon, 1984.

National Council of Teachers of Mathematics. *An Agenda for Action.* Reston, Va.: The Council, 1980.

CAN YOUR ALGEBRA CLASS SOLVE THIS?

Problem 11. The equation $ax^2 - 2x\sqrt{2} + c = 0$ has real constants a and c with a zero discriminant. Describe the roots.

Solution on page 248

11

Polynomials in the School Curriculum

Theodore Eisenberg
Tommy Dreyfus

O VER the past twenty years there seems to have been a definite de-emphasis on topics surrounding polynomials in the school curriculum. This can be witnessed even at the simplest level of factoring a quadratic expression to obtain its roots (Hartzler 1982; Nicely 1985). In Israel, for example, students are initially taught to find the roots to quadratic equations by factoring and then setting each factor equal to zero. Soon thereafter the quadratic formula is developed. The formula becomes so ingrained in students, especially students of lesser mathematics ability, that it becomes a mechanical procedure, an algorithm void of meaning that is used indiscriminately to solve all quadratic equations. For example, in a large lecture class in introductory calculus for nonscience majors at Ben-Gurion University, fully 50 percent of the students solved equations like $x^2 + 5x = 0$ and $10x - x^2 = 0$ with the quadratic formula! For equations like $x^2 - 6x + 5 = 0$ or $2x^2 - 3x - 2 = 0$ this percentage rose to 70 percent. Moreover, most students had no idea of how to attack the problem of finding the roots of higher-order equations. Indeed, they did not understand what it means to "find the roots of an equation." To them, solving a quadratic equation meant putting numbers into a formula.

On another occasion, twenty-five in-service junior and senior high school mathematics teachers were asked to develop the formulas for the sum and product of the roots to the general quadratic equation in a single variable. Only two of them gave a factoring argument. Twenty obtained the correct results by adding and multiplying the two branches of the quadratic formula. They also mentioned that this is how they teach it in the classroom. The remaining three did not respond at all. Moreover, not a single one saw how these formulas can be generalized to polynomial equations of higher degree. Indeed, it is impossible to see this generalization when approaching the problem by manipulating the quadratic formula. The point is that students seem to be learning techniques at the expense of understanding the larger picture.

The outlook of both students and teachers in the surveys above is extremely limited. The trend to emphasize procedures rather than the underlying structures (Coopersmith 1984) not only carries with it the danger of considering mathematics as simply a compendium of algorithms but also makes it very difficult for students to generalize and apply what they have learned. This hardly seems to be the direction in which we want to be moving. The use of the quadratic formula must be de-emphasized; rather, quadratic equations should be explicitly considered as second-degree polynomial equations, and as such, constitute examples for more general polynomials. The underpinnings of this are accessible to all students and are very important for them. The sort of reasoning patterns one develops in working with polynomial equations generalizes to other settings. Higher-level notions of functions can be introduced through polynomials. The solution to problems that when initially encountered seem to have no connection to polynomials do, in the end, depend heavily on them. Polynomials are ubiquitous in mathematics, and it is important that students be well grounded in them. The purpose of this article is to give a glimpse of how and why polynomials should be part of the school curriculum.

FACTORING POLYNOMIALS

If r is a root of the polynomial $f(x) = a_0 x^n + a_1 x^{n-1} + \ldots + a_{n-1} x + a_n$, then $(x - r)$ is a factor of $f(x)$. That is, there is a polynomial $g(x)$ such that $(x - r)g(x) = f(x)$. Similarly, if $f(x) = (x - r)g(x)$, then r is a root of $f(x)$. This is the factor theorem. Clearly, the degree of $g(x)$ is one less than that of $f(x)$.

The result that r is a root, if and only if $(x - r)$ is a factor of the polynomial, is easily grasped by the students, but they initially encounter problems in constructing the polynomial $g(x)$. Here is a perfect opportunity to discuss long division of polynomials by modeling the discussion against long division of real numbers (another soon-to-be-obsolete activity? [Usiskin 1983]). Using this analogy, even weak algebra students have little trouble grasping and intelligently discussing the idea of the remainder $r(x)$ in $f(x) = h(x)g(x) + r(x)$. Moreover, because of this modeling procedure, students are encouraged to search for arithmetic similarities for other algebraic activities. Research indicates that such similarities are efficient devices in understanding algebraic structures (Dreyfus and Thompson 1985).

If factoring is stressed and often used, as we think it should be, then there is no reason to stick with the long-division algorithm forever. It is worthwhile to invest the little time that is needed to understand synthetic division, which seems to be another topic that is simply not taught any more. In an informal survey of ten newly published algebra textbooks designed for advanced high school and junior college students, only half contained the synthetic division algorithm, and even then it was seldom emphasized. But even those students who know synthetic division seem to retain only the mechanics of the procedure. Interviews with high school mathematics teachers and computer science students who were familiar with the algorithm revealed that even with probing, they were unable to construct an acceptable argument for why it works.

But why is it important to factor? The remainder of this article may be considered an extended answer to this question. The answer begins with a rather obvious aspect, modeled on the real numbers. If the aim is to find the roots of the polynomial $f(x)$, and if it is possible to factor $f(x)$ into two factors $g(x)$ and $h(x)$, then $g(x) = 0$ or $h(x) = 0$. Hence the problem is now to find the roots of $g(x)$ and of $h(x)$, each of which is of degree less than that of $f(x)$. This method generalizes to many other types of problems, such as finding the roots to $\sin 3x + 3 \sin 2x + 3 \sin x = 0$, proving that 80 divides $3^{2n} - 1$ for all even numbers n, or proving that ABC is a right triangle if $\sin A \sin B = \cos C$. The method of trying to factor $f(x)$ when faced with the equation $f(x) = 0$ should be so ingrained in the students that it should *always* be the first method they consider. One way to achieve this is to present a considerable number of standard exercises in factoring polynomials and to follow them up with some harder exercises—for example, to find all polynomials such that $(x - 1)f(x + 1) - (x + 2)f(x) = 0$ or to prove that a quadrilateral inscribed in a circle with sides whose lengths are consecutive integers cannot have an integer area (Garfunkel 1985). The solutions to these problems depend heavily on factoring techniques and tie together many of the concepts on polynomials that at present are only studied in isolated form, if at all.

RATIONAL ROOTS

It is immediately evident to students that every polynomial equation $f(x) = 0$ with rational coefficients can be turned into one with integer coefficients. The question naturally arises as to the number of zeros of a polynomial. This question is not well posed as long as the domain in which solutions are sought is not specified. For instance, the cubic $x^3 - x^2 + x - 1$ has only one integer zero ($x = 1$) but two more complex zeros ($x = i$ and $x = -i$). No general results are known about the number of integer or real zeros of polynomials, but it is known that every polynomial of degree n has at most n distinct roots. This is the fundamental theorem of algebra.

Frequently, one has to find one root of a given polynomial; this happens, for example, when faced with solving a third-order linear differential equation. Such an equation may be solved by finding the roots of its characteristic polynomial. This is a polynomial associated with the differential equation; if the equation is of order n, then the polynomial will be of degree n, that is, of degree 3 in our case. One therefore needs to start by finding a root of this third-degree polynomial (the other two roots are then easy to find). A theorem enabling one to do this in many cases, the rational roots theorem, has been dropped from the majority of high school textbooks. Several of its applications will be given below.

THEOREM. *If the polynomial equation* $a_0 x^n + a_1 x^{n-1} + \ldots + a_{n-1} x + a_n = 0$ *with integer coefficients has a rational root of the form p/q (where p/q is in reduced form, i.e., p and q have no common divisors: $[p,q] = 1$), then p divides a_n and q divides a_0.*

This theorem is proved by substituting the root $x = p/q$ into the equation, multiplying through by q^n, and exploiting the fact that p (and q) must divide the resulting expression.

It is exactly the rational roots theorem that allows one to focus on the first and last terms of a trinomial when wanting to write it as the product of two binomials. The usual method for factoring a trinomial of the form $ax^2 + bx + c$ into the product of two binomials $(rx + m)(sx + n)$ focuses on finding the numbers $r, s, m,$ and n such that $rs = a, mn = c,$ and $rn + sm = b$. But this is an application of the theorem above because $r(x + m/r)s(x + n/s) = 0$ implies that $-m/r$ and $-n/s$ are rational roots.

The theorem also gives a method for conclusively determining whether or not a polynomial equation has a rational root: simply form all possible quotients from the factors of a_n and a_0 and check whether or not they are roots. For example, in order to find the roots of $f(x) = 24x^3 - 46x^2 + 29x - 6$, form the list of all ratios of the factors of 6 divided by those of 24. Noting that $f(0) < 0$ and $f(1) > 0$ provides a judicious way to enter the list, namely, focusing on the candidates between 0 and 1. Thus $x = 1/2$ is identified as a root. The other two roots are easily found by factoring the polynomial. The list also provides a systematic way for *all* rational roots of a given polynomial.

The power of the rational roots theorem again becomes apparent in graphing polynomials (see "Graphing") and when considering irreducible polynomials ("Irreducible Polynomials").

SUMS AND PRODUCTS OF ROOTS

Many students and teachers develop the usual formulas for the sum and product of the roots to the quadratic equation by manipulating the quadratic formula. This method is certainly valid, but there is a more powerful way, which generalizes to higher-order polynomials by modeling a derived equation against the original polynomial. Again, this more powerful method is based on factoring the polynomial under consideration. Start with a quadratic equation $ax^2 + bx + c = 0$ with roots r_1 and r_2:

$$ax^2 + bx + c = a(x - r_1)(x - r_2) = a[x^2 - (r_1 + r_2)x + r_1r_2] = 0.$$

Dividing by a and comparing coefficients gives the usual formulas, $(r_1 + r_2) = -b/a$ and $r_1r_2 = c/a$.

This is easily generalized to polynomials of higher degree. For example, $a_0x^5 + a_1x^4 + a_2x^3 + a_3x^2 + a_4x + a_5 = a_0(x - r_1)(x - r_2)(x - r_3)(x - r_4) \cdot (x - r_5) = 0$ leads to

$$\sum_{i=1}^{5} r_i = -\frac{a_1}{a_0} \qquad \sum_{i=j} r_ir_j = \frac{a_2}{a_0} \qquad r_1r_2r_3r_4r_5 = -\frac{a_5}{a_0}.$$

Two interesting exercises can be done here: (1) If $a_0x^3 + a_1x^2 + a_2x + a_3 = 0$ has roots r_1, r_2, and r_3, find an equation that will have roots kr_1, kr_2, and kr_3 where k is any real (or complex) number. (2) Find the relationship between the roots of the polynomial $f(x) = a_0x^n + a_1x^{n-1} + \ldots + a_{n-1}x + a_n$ and those of the polynomial with the order of the coefficients reversed: $f(x) = a_nx^n + a_{n-1}x^{n-1} + \ldots + a_1x + a_0$ (Schoenfeld 1983). There is absolutely no reason why this sort of thing should not be in the curriculum, for it gives insight into the structure of polynomials themselves.

GRAPHING

Another place in the curriculum where factoring is useful is in graphing polynomials and rational functions. When graphing a polynomial, one of the first questions one should ask is whether or not it has rational roots. The easiest way to determine this is to use the rational roots theorem.

Moreover, synthetic division can be used in finding points on the graph of $f(x)$, all of which have the same y coordinate (Olson and Goff 1985). This is particularly nice with cubic polynomials. For example, let $f(x) = 8x^3 - 58x^2 - x + 111$. Dividing $f(x)$ by $x - 7$ yields $f(x) = (x - 7)(8x^2 - 2x - 15) + 6$. Although 7 is not a root, the point $(7,6)$ is on the graph of $f(x)$. But more than that, $8x^2 - 2x - 15$ has two roots, $-5/4$ and $3/2$, so $(-5/4,6)$ and $(3/2,6)$ are also points on the graph of $f(x)$. In general, if $f(x) = (x - a)q(x) + r$, then (a,r) is a point on the graph because $f(a) = r$. But if s is a root of $q(x)$, then (s,r) is also a point on the graph. Hence focusing on the roots of $q(x)$ gives points on the graph with the same y-coordinate.

Another place in the high school curriculum where factoring becomes useful is in graphing rational functions $f(x) = p(x)/q(x)$ where $p(x)$ and $q(x)$ are polynomials. Such graphing exercises force one to apply the theory of polynomials. When graphing

$$f(x) = \frac{x^4 + 4x^3 + 5x^2 + 8x + 6}{x^2 + 3x - 4},$$

one must factor the denominator in order to obtain the vertical asymptotes. Then long division can be used to obtain the asymptotic parabola, which will guide the graph of the function as $x \to \infty$. It is certainly possible to get a feel for the graph without using tools of calculus (see fig. 11.1).

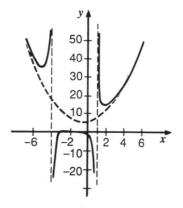

Fig. 11.1 Graph of a rational function

Exercises like this are extremely important for the student to understand (Hornsby and Cole 1986). They build on skills previously learned and enable students to look at functions holistically, focusing on their general behavior rather than minute details. The basic building blocks for obtaining this skill are contained in the properties of polynomials.

Furthermore, by studying polynomials when $(x - a)$ has been substituted for x, one sees how the graph of $f(x - a)$ is related to the original graph of $f(x)$; the graphs of $f(x) + c$, $f(kx)$, and $kf(x)$ can be investigated in a similar manner. These four substitutions correspond to the transformations of sliding and stretching (or shrinking) the original graph. Apart from their intrinsic interest, they are important in applications such as the function $f(x) = A \cos(\omega x + \phi)$, which occurs in the description of periodic phenomena in the sciences. Again, embedded in polynomials are skills that then turn out to have more general validity.

Students often need to know how to build a polynomial whose graph passes through a prespecified finite set of points in the plane, a typical problem encountered in physics and chemistry laboratories. There are several ways of handling such problems, but the most efficient method is Lagrange interpolation (Flanders, Korfhage, and Price 1970). The polynomial

$$f(x) = \frac{(x - x_2)(x - x_3)}{(x_1 - x_2)(x_1 - x_3)}y_1 + \frac{(x - x_1)(x - x_3)}{(x_2 - x_1)(x_2 - x_3)}y_2 + \frac{(x - x_1)(x - x_2)}{(x_3 - x_1)(x_3 - x_2)}y_3$$

passes through the points (x_1, y_1), (x_2, y_2), and (x_3, y_3). Constructing the various segments of the polynomials employs many of the principles discussed above.

IRREDUCIBLE POLYNOMIALS

If $x^3 + px^2 + qx + r = 0$, where p, q, and r are rational numbers, cannot be factored over the rational numbers, then its roots cannot be constructed with Euclidean tools. This theorem is the key in proving that with Euclidean tools, a given angle cannot be trisected, a cube cannot be doubled, a circle cannot be squared, and a regular 7-gon cannot be constructed (Bold 1982; Courant and Robbins 1941). The tie between polynomials and the geometric construction problems of antiquity is strong and provides another good reason why polynomials should be part of the school curriculum.

Other instructive topics with polynomials are Descartes' rule of signs, which gives the maximal number of positive and negative roots of a polynomial; De Moivre's theorem, which aids in finding all roots of $x^n = 1$; and Newton's method for approximating roots. Many aspects of mathematical thought can be learned by studying polynomials. They are important in the curriculum, and they should remain there.

REFERENCES

Bold, Benjamin. *Famous Problems of Geometry and How to Solve Them.* New York: Dover Publications, 1982.

Coopersmith, Art. "Factoring Trinomials: Trial and Error? Hardly Ever!" *Mathematics Teacher* 77 (1984): 194–95.

Courant, Richard, and Herbert E. Robbins. *What Is Mathematics?* London: Oxford University Press, 1941.

Dreyfus, Tommy, and Patrick W. Thompson. "Microworlds and van Hiele Levels." In *Proceedings of the Ninth International Conference for the Psychology of Mathematics Education*, edited by L. Streefland, pp. 5–11. Utrecht, The Netherlands: State University, 1985.

Flanders, Harley, Robert R. Korfhage, and Justin J. Price. *Calculus.* New York: Academic Press, 1970.

Garfunkel, Jack. "Problem No. 4072." *School Science and Mathematics* 85 (December 1985): 714.

Hartzler, Stanley J. "Secondary Algebra Textbooks in the United States 1806–1982: Selected Descriptive and Historical Trends." (Doctoral dissertation, University of Texas at Austin, 1982) *Dissertation Abstracts International* 43 (1982): 2265A.

Hornsby, E. John, Jr., and Jeffery A. Cole. "Rational Functions: Ignored Too Long in the High School Curriculum." *Mathematics Teacher* 79 (December 1986): 691–98.

Nicely, Robert F. "Higher Order Thinking Skills in Mathematics Textbooks." *Educational Leadership* 42 (March 1985): 26–30.

Olson, Melfried, and Gerald K. Goff. "Applications of Synthetic Division to Graphing Cubic Equations." *School Science and Mathematics* 85 (November 1985): 591–94.

Schoenfeld, Alan H. *Problem Solving in the Mathematics Curriculum—a Report, Recommendations and an Annotated Bibliography.* Washington, D.C.: Mathematical Association of America, 1983.

Usiskin, Zalman. "Arithmetic in a Calculator Age." *Arithmetic Teacher* 30 (December 1983): 2.

12

Teaching Elementary Algebra with a Word Problem Focus

Harold L. Schoen

- One side of a right triangular piece of property is 50 units long. Parallel to the other side and 20 units from the other side a line is drawn cutting off a right trapezoidal area of 5.20 units. Find the lengths of the parallel sides of this trapezoid. (Babylonian, ca. 3000 B.C. [Boyer 1968, p. 47])

- If one pipe can fill a cistern in one day, a second in two days, a third in three days, and a fourth in four days, how long will it take all four running together to fill it? (Greek, ca. A.D. 520 [Boyer 1968, p. 214])

FROM the beginning of written history, people have been interested in applying mathematics to situations that are described verbally. In the twentieth century, too, with a twenty-year hiatus in the 1960s and 1970s, an important goal of school mathematics has been the solution of applications and word problems. Recently mathematics educators have turned a great deal of attention to this goal, especially since the appearance of the NCTM's (1980) *Agenda for Action*, which declared that problem solving should be the focus of mathematics instruction. The authors of the *Agenda* had a broad view of problem solving in mind, one that excluded rote, algorithmic skills but that included, among other things, standard textbook word problems and other applications of mathematics.

The thesis of this article is that it is possible to focus on interesting applications and word problems in the teaching of first-year algebra without deleting important topics. The ideas presented here are based to a large extent on a four-year project at the University of Iowa that implemented such an approach in elementary algebra, a remedial college course. The algebraic topics in the course are about the same as those in first-year algebra. A component of the course was a tutorial laboratory that provided one-to-one tutoring for thousands of students over the duration of the project. In this way, much was discovered about how these students think about algebra and about word problems, and these discoveries were then used in revisions of the course. Most of the ideas in the article have been presented to groups of secondary school teachers, who judged them to be within the realm of practicality for first-year algebra classes. The algebraic concepts and procedures taught in the course and used as examples in this article are, for the most part, the traditional ones. However, it seems likely that the philosophy underlying the teaching approach could be adjusted for other less traditional content in first-year algebra or for courses at higher levels.

GENERAL TEACHING PRINCIPLES

Principles for effective mathematics teaching such as high-quality teacher-student interaction, clarity of explanations, much on-task time, and regular review are important in any classroom and will not be discussed here. However, the first two of the recommendations that follow are based on general teaching principles that are especially important for teaching an algebra course focused on solving word problems.

Recommendation 1

Build new learning on students' existing knowledge and understanding.

Students' minds are not blank slates when they enter first-year algebra. Students not only possess knowledge from previous mathematics courses but also have many beliefs and preconceptions about algebra, about word problems, and about the real-world concepts described in word problems. For example, recent work with area formulas is likely to be fresh in students' minds, yet conceptual understanding of area may be lacking. Revisiting the idea of area with a conceptual and applied emphasis prior to formal algebra work has several potential payoffs. First, to the students it is a "concrete" use of letters to represent numbers and thus a good start on the general concept of variable. Second, the concept of area and the associated formulas will be needed in many algebraic word problems later in the course. Third, geometric measurement provides a good source for early, prealgebra problem-solving and application experiences. The following is an example of a real-world application that requires some conceptual understanding of area.

- Carol's rectangular yard measures 9 m long and 5.5 m wide. She decides to put two circular flower gardens in her yard, each with a radius of 0.75 m. The remainder of her yard will be a vegetable garden. How many square meters of her yard will be used for flowers? For vegetables? Can Carol arrange the flower gardens to allow a 3 m × 7 m rectangular region for a potato patch? Why or why not?

Other topics that should be revisited early are ratio, proportion, and percent, since they are needed to solve many algebraic applications. In spite of previous instruction, some students will not have a good understanding of these topics. In fact, past instruction may interfere with algebraic approaches, as in the following example:

- An 80% acid solution is mixed with an 18% acid solution to get 3 gallons of solution that is one-third acid. How much of each solution must be used?

A typical algebraic strategy is to assign x to the amount of 80% solution and y to the amount of 18% solution. This system of equations is then an appropriate model.

$$x + y = 3$$
$$(80\%)x + (18\%)y = (1/3)(3)$$

The student who has seen percent treated only as a proportion (i.e., 18% of 50 is thought of as $18/100 = n/50$) will not have the facility to set up the second equation, in which 80% of x must be viewed as 80% *times* x.

Recommendation 2
Lead gradually from verbalization to algebraic symbolism.

This recommendation is related to the previous one, in the sense that most students are able to understand written or spoken English when they enter first-year algebra. We should capitalize on this ability by writing and speaking in English about the mathematical ideas prior to, and along with, the introduction of the symbolism. It is no accident that the historical development of algebraic symbolism began with a period of rhetorical or verbal algebra at least three millennia in duration. The rhetorical period was followed by more than a thousand years in which algebraic discourse gradually moved from the rhetorical to the symbolic.

Thus, the algebraic symbolism we teach high school students is a phenomenon of the last five hundred years and the product of a gradual development over a period ten times that long. To rush students into algebraic symbolism too quickly ignores the need for the verbal grounding and gradual symbolization that is suggested by history and supported by research in the teaching and learning of algebra.

For example, consider the multiplication property for equality in both a verbal and a symbolic form.

Verbal: Given an equation, we can multiply both sides by the same number and the sides will still be equal.

Symbolic: For real numbers a, b, and c, if $a = b$, then $ca = cb$.

Now, suppose a student needs to apply this property to clear the following equation of fractions:

$$(1/4)x - (3/8) = (3/4)x$$

To apply the symbolic form, a must be thought of as $(1/4)x - (3/8)$, b as $(3/4)x$, and c as 8. This is a highly complex use of the variable idea, which students at this level are just struggling to learn in its simpler forms. The verbal form of the property, however, can be applied easily and meaningfully to the equation.

INTEGRATING WORD PROBLEMS WITH OTHER ALGEBRAIC CONTENT

If applications are to be taken seriously in algebra, they must be an important part of the content, not just afterthoughts at the ends of chapters. Each of the following recommendations will help to accomplish the kind of integration that is needed.

Recommendation 3
Introduce algebraic topics with applications.

Real-world situations can be used to establish a need for many topics in algebra. In a sense, this reverses the typical sequence of teaching algebraic concepts and procedures before applications. For example, subtraction of negative integers can be introduced with this scenario:

- After three hands in a card game, Phil was 8 points "in the hole." In other words, his score was -8. Before the next hand, Phil discovered that the scorer had made a mistake and had given him 12 negative points too many. Phil found his corrected score by taking away (that is, subtracting) the -12 points. He subtracted $(-8) - (-12)$. What is Phil's corrected score?

Trying to approach every topic in first-year algebra with a word problem would be abusing a good idea. However, when used for the many appropriate topics, this practice will emphasize the importance of word problems, provide an initial connection between the concept or procedure and its uses, and serve as another exposure to verbal applications.

Recommendation 4

Teach algebraic topics from the perspective of how they can be applied.

In addition to introducing topics with word problems, teachers can use applications as embodiments of algebraic concepts. As such, they are tools for teaching the concept itself, not just for initially introducing it or for finally applying it. A good example is graphing linear equations in two variables followed by systems of two equations in two unknowns. The following is one of a myriad of verbally described real-world situations that can be modeled by a linear equation and used to introduce and teach this sequence of topics.

- It costs $7.50 to prepare the type for a certain pamphlet and $0.25 a copy for production.

  1. Make a table that gives the total cost y of producing x pamphlets where $x = 10, 20, 30, 40$, and 50.
  2. Graph the points in step 1 on an xy coordinate system.
  3. Write an equation that gives y in terms of x.
  4. Graph the equation in step 3.
  5. What is the total cost of producing 500 pamphlets?

This application is a good vehicle for analyzing and discussing the concrete meanings of domain and range, since both are discrete and bounded in this example. The coordinates of points on the graph, the slope, and the y-intercept of the continuous extension of this cost function take on real meanings as well. Furthermore, systems of equations can now be introduced in a natural way by adding a second condition to the application above, such as the following:

- The pamphlets above are sold for $0.75 each.

  1. Write an equation that gives the total income y from selling x pamphlets.
  2. Graph the equation in step 1 on the same set of axes as you did the cost equation.
  3. How many pamphlets must be sold to break even, that is, for cost to equal income?

As this example illustrates, the potential of word problems as tools in teaching algebraic topics in a meaningful way should not be underestimated.

Recommendation 5

Teach and model specific heuristics as aids in understanding and solving word problems.

Students need to learn strategies for solving word problems. Using tables, diagrams, formulas, and graphs; identifying the wanted and given; translating English phrases to algebraic symbols; and checking answers with the conditions of the problem are all useful strategies to emphasize throughout the course. When the instruction is focused on solving word problems, there seems to be little need to take a block of time to teach these heuristics. Since many heuristics are modeled and used regularly, students quickly become comfortable with them.

In addition to teaching general heuristics, it is important to emphasize the mathematical structure of a given word problem in order to help students see why an equation or system of equations is a good model. One strategy for doing this is to have students initially write in English the equation that models the problem, as in the example below.

- The depreciation on an automobile during the first year is 20% of the initial cost. After that the yearly depreciation is 10% of the initial cost. Suppose a car has depreciated $3000 in 5 years. What was the initial cost of the car?

The usual approach to solving this problem calls for a translation of parts of the problem; for example, let x be the initial cost and then $0.2x$ is the first year's depreciation. This focuses the student's attention on parts of the problem, drawing it away from a consideration of the structure of the whole problem. Requiring students to write a verbal equation such as the following, however, keeps their attention on the problem's structure:

(depreciation after 5 years) = (depreciation after 1 year)
                    + 4 × (depreciation each year thereafter)

This practice can be dropped once students develop good facility within a domain of applications.

Recommendation 6

Hold students accountable for solving word problems.

Even if teachers make an attempt to teach word problems as applications of the recently learned algebraic techniques, students still tend to see word problems as peripheral to algebra. Many college students in our remedial elementary algebra course said that they never tried to learn how to solve word problems in high school algebra. In a few instances their teachers skipped word problems entirely, but more often the students simply decided not to try to understand them. In their judgment, they might never be able to understand word problems, but they had a fair chance of learning most of the algorithmic algebraic techniques. They also noted that they could miss all the points on word problems on any test and still get a grade of B. Hence, they considered it good strategy to spend their time learning the algebraic techniques and to ignore the word problems.

To counteract such avoidance, 40 percent or more of the grade in the course focused on solving word problems should depend on the student's ability to analyze and solve verbal applications. There are many ways to test this ability. One way is to pose a single application and ask many questions about it, aimed at testing various analysis and problem-solving skills. This approach results in several test items from a single word problem, thus saving time for both the test writer and the test taker. Such test items should be scored so that errors on one question do not result in penalties on subsequent questions. It is also important to note that applications usually require thinking at a higher level than skill items, so do not expect all students to score 90 percent or even 70 percent on application tests. Adjust your grading system accordingly.

IMPLEMENTING THE RECOMMENDATIONS

In order to make word problem solving the focus of an algebra course, it is necessary but not sufficient to include many good word problems. It is also important that the word problems be integrated with the other algebraic content and that the instruction be designed to help students develop the skills needed to solve the word problems, not just to master the algebraic techniques. Finally, students must be held accountable for being able to solve word problems. On the basis of our project at the University of Iowa, such a course is possible without slighting important algebraic skills.

To implement this teaching approach does not require many changes in algebraic content but rather changes in emphasis and point of view. Except for the review of geometric measurement, ratio, proportion, and percent discussed following Recommendation 1, the content of our course comprised the traditional topics. The time spent on this review was obtained by de-emphasizing the symbolically stated number properties. Recommendations 2 and 6 are directly under the control of the teacher. Teachers can simply plan to use a more verbal approach and to make a significant part of the course grade depend on students' ability to solve word problems. Recommendations 3 and 4, which concern the introduction of algebraic topics with word problems and presenting topics from the perspective of how they are applied, respectively, require a change in the placement of the word problems. If a textbook has good word problems at the end of a chapter, problems can be chosen from them to introduce and present along with the algebraic topics earlier in the chapter.

Of course, the teacher's job will be easier if the textbook contains good word problems along with ideas for using them to introduce and teach the algebraic topics. Some newer textbooks are beginning to include more word problems, but regardless of the textbook's emphasis, a few sources of good applications will prove to be extremely valuable. A particularly good teacher resource is the sourcebook prepared by a joint committee of the MAA and NCTM and available from the NCTM (MAA/NCTM 1980).

Although a word-problem in teaching first-year algebra is possible and even practical, it will surely mean that more class time must be spent on word problems. In our course, we borrowed from time previously spent on the drill and practice of algebraic skills for their own sake. We found that students did not profit from the large amounts of time that had been spent on skill practice, but that some practice along with the review and reinforcement of skills that arose naturally in solving the word problems was sufficient to maintain acceptable skill levels. In fact, our word-problem focus resulted in a more effective, interesting, and useful algebra course.

REFERENCES

Boyer, Carl B. *A History of Mathematics*. New York: John Wiley & Sons, 1968.

Mathematical Association of America and National Council of Teachers of Mathematics. *A Sourcebook of Applications of School Mathematics*. Reston, Va.: NCTM, 1980.

National Council of Teachers of Mathematics. *An Agenda for Action*. Reston, Va.: NCTM, 1980.

CAN YOUR ALGEBRA CLASS SOLVE THIS?

Problem 12. A set of n numbers has the sum s. Each number of the set is increased by 20, then multiplied by 5, and then decreased by 20. What is the sum of the numbers in the new set in terms of s and n?

Solution on page 248

13

From Words to Algebra: Mending Misconceptions

Jack Lochhead
José P. Mestre

R ECENT research indicates that many students appear to have inordinate difficulties solving certain types of fairly simple algebraic word problems, particularly in regard to translating written language to mathematical language. In problems where students are asked to read a sentence stating a relationship between two variables and then write an equation expressing that relationship, they frequently write the reverse of what they intend (Clement, Lochhead, and Monk 1981). For example, in the problem below, 37 percent of college engineering students answered incorrectly:

> Write an equation using the variables S and P to represent the following statement: There are six times as many students as professors at this university. Use S for the number of students and P for the number of professors.

Two-thirds of those who answered incorrectly chose the answer $6S = P$, where the variables are reversed.

THE PROBLEM

The first point to note in the results above is that many of the errors follow a consistent pattern; they do not appear to be random. Interviews with more than twenty students who made the error showed that few, if any, of the errors stemmed from a misreading of the problem (Clement, Lochhead, and

The work reported herein was funded in part by grants from the National Science Foundation, the National Institute of Education, the Exxon Education Foundation, the Fund for the Improvement of Post Secondary Education, and Digital Equipment Corporation. The views expressed are those of the authors and do not necessarily reflect the position, policy, or endorsement of these organizations.

Monk 1981). Not one student interviewed indicated that there were more professors than students. The source of the error stems from misconceptions concerning the structure and interpretation of algebraic statements and of the process by which one translates between written language and algebraic language.

It is also important to emphasize that the difficulty we are discussing stems neither from a lack of algebraic fluency (in the sense of symbol manipulation) nor from an inability to read. Figure 13.1 shows that 95 percent of the students sampled could perform algebraic manipulations (problem 1) and that they could correctly read and solve simple word problems (problem 2); however, they could not correctly translate the "students and professors" problem (problem 3) or the "cheesecakes and strudels" problem (problem 4) into mathematical language.

In addition, figure 13.1 indicates that the difficulties are more extensive than one might expect. Students have inordinate difficulties in three other classes of problems: (1) problems where they are asked to write an equation to represent the relationship between two variables given in tabular form (problem 5); (2) problems where they are asked to write a sentence given a simple two-variable linear equation of the type we have been discussing (problem 6); and (3) problems where they are asked to write an equation to

---

Test Questions ($n = 150$)

	Correct Answer	Percent Correct	Typical Wrong Answer
1. Solve for x: $\dfrac{6}{4} = \dfrac{30}{x}$	$x = 20$	95%	
2. Jones sometimes goes to visit his friend Lubboft driving 60 miles and using 3 gallons of gas. When he visits his friend Schwartz, he drives 90 miles and uses ? gallons of gas. (Assume the same driving conditions in both cases.)	4 1/2	93%	
3. Write an equation using the variables S and P to represent the following statement: "There are six times as many students as professors at this university." Use S for the number of students and P for the number of professors.	$S = 6P$	63%	$6S = P$
4. Write an equation using the variables C and S to represent the following statement: "At Mindy's restaurant, for every four people who order cheesecake, there are five people who order strudel." Let C represent the number of cheesecakes and S represent the number of strudels ordered.	$5C = 4S$	27%	$4C = 5S$

Fig. 13.1

	No. of Students Tested	Percent Correct

5. Weights are hung on the end of a spring and the stretch of the spring is measured. The data are shown in the table below: 381   42%

Stretch S (cm)	Weight W (g)
3	100
6	200
9	300
12	400

Write an equation that will allow you to predict the stretch (S) given the weight (W).

6. Write a sentence in English that gives the same information as the following equation: "$A = 7S$." "A" is the number of assemblers in a factory. "S" is the number of solderers in a factory.   34   29%

7. A man takes a photograph from an airplane of some of the cows and pigs in a large field full of cows and pigs. He is sure that he has photographed a typical sample of the animals in the field. Write an equation using the letters C and P to describe the relationship between C, the number of cows, and P, the number of pigs, in the field. The equation should allow you to calculate the number of cows if given the number of pigs.   85   38%

photograph

Fig. 13.1 (cont.)

represent the relationship between two variables given in pictorial form (problem 7).

The sources of these errors have been discussed extensively in the research literature (Clement 1982; Clement, Lochhead, and Monk 1981; Mestre, Gerace, and Lochhead 1982; Rosnick 1981; Rosnick and Clement 1980; Soloway, Lochhead, and Clement 1982). We shall restrict our discussion to two types of errors that are particularly salient. First, students exhibit a strong proclivity toward performing a left-to-right word-order match when they translate the "students and professors" and the "cheesecakes and strudels" problems. This leads to expressing "six times as many students" as "$6S$," and "professors" as "P," and thus the common error, $6S = P$.

Second, students often confuse the distinction between variables and labels. The symbols "S" and "P" are often interpreted as labels for "students" and "professors," rather than as variables to represent the "number of students" and the "number of professors"; this leads to interpreting $6S = P$ to mean "six students for every one professor" in a way similar to interpreting 3 ft = 1 yd to mean "three feet for every one yard." The variable-label confusion is also the main mechanism causing the large error rate in problem 7 of the figure, in which an equation must be written from pictorial data; here, $5C$ and P are very convenient labels for the five cows and one pig in the picture, and thus students write the equation $5C = P$.

What is terrifying about these data is that the students sampled were mathematics, science, and engineering majors enrolled in some fairly prestigious universities across the country. We would like to think that people who will be building our future bridges, computers, and airplanes would at least be able to perform flawlessly in the simple algebraic problems of figure 13.1. What is most lamentable to us is that our educational system does not seem to address the conceptual issues that would help students overcome these types of misconceptions.

Although these algebraic misconceptions were first observed among college students, evidence now shows that they are widespread across both age and nationality. For example, the same error patterns were observed in a study with ninth graders enrolled in algebra 1 although with a much higher incidence (Mestre and Gerace 1986). Further, similar errors have been observed with Hispanic students (Mestre, Gerace, and Lochhead 1982) as well as with students from Fiji, Israel, and Japan (Lochhead, Eylon, Ikeda, and Kishor 1985; Mestre and Lochhead 1983).

It therefore appears that instruction in algebra in the United States, Israel, and Fiji—and, we expect, just about everywhere else—does not provide students with an adequate opportunity to learn how to interpret mathematical symbol strings. Students do not learn to read and write in mathematics! This omission not only limits their performance on word problems but also places them at a severe disadvantage when it comes to learning the symbol manipulation rules of algebra. Without the ability to interpret expressions, students have no mechanism for verifying whether a particular procedure is correct. Thus, they often have to rely on rote recall to solve problems.

SOLUTIONS

Although many mathematics teachers are familiar with the general translation difficulty, most are perplexed by its persistence. We have studied it intensely for nearly a decade but remain astonished at how difficult it is to teach students to avoid the common pitfalls. In the remainder of this article we discuss a few practical techniques that have been of use in our own instructional efforts.

The most direct technique is to start simply, providing students with ample practice at the translation process itself, isolated from all other aspects of problem solving. Whimbey and Lochhead (1981) have a series of simple exercises (see fig. 13.2) that gradually introduce students to a few of the common difficulties. However, it is a mistake to think that careful preparation can protect them from all the traps. A certain amount of confusion among the different representational systems used in mathematics is inherent to the system and thus unavoidable. Full mastery is possible only after students have struggled with the confusion and begun to understand its origin. Narode and Lochhead (1985) suggest the following problem:

> Write an equation which can be used to calculate the number of feet in a measurement given the number of yards. Use the letters "F" for the number of feet and "Y" for the number of yards.

Sample Problems to Help Students with Translation Difficulties

Translate each English sentence into mathematical language (ML). Use letters which are abbreviations for words. Let F stand for "Fred's income," let h stand for "number of hours," and so on. Some of the sentences have already been translated as illustrations.

1. The combined incomes of Fred and Harry equal $490.

   Math language: $F + H = 490$

12. A man worked for 20 hours at $3 an hour plus 10 hours at $5 an hour for a total of $110.

    ML: $(20)(3) + (10)(5) = 110$

24. Larry is four times as old as his son Bobby.

    ML:

    The correct answer is $L = 4B$. If you got this problem wrong, try first to rewrite the English sentence so that it includes the word "equals." For example:
    Larry's age equals four times Bobby's age.

37. Ten pounds less than 50 pounds equals 40 pounds.

    ML: (The answer is not $10 - 50 = 40$.)

45. Four more than five times a number equals one less than six times that number.

    ML: (The answer is not $5n + 4 = 1 - 6n$.)

192. A man worked a certain number of hours at $3 an hour and the same number of hours plus 20 more hours at $4 an hour, earning a total of $150.

     ML: (This is *not* correct: $3n + n = 20(4) = 150$.)

Fig. 13.2. A sampling of problems from *Developing Mathematical Skills* (chap. 10) by Whimbey and Lochhead, 1981.

Obviously this is a setup, designed to create confusion between the correct answer $3Y = F$ and the units convention 1 yd = 3 ft. The purpose of the problem is to get students thoroughly confused as preparation for bringing them to the beginning of an understanding of the difference between variables (Y and F) and labels (yd and ft). Unless students become completely exasperated by their confusion, they are apt to consider the distinction between variables and labels mere pedantic quibbling.

Rosnick (1982) also confronts potential confusion head-on with this problem:

> I went to the store and bought the same number of books as records. Books cost two dollars each and records cost six dollars each. I spent $40 altogether. Assuming that the equation $2B + 6R = 40$ is correct, what is wrong, if anything, with the following reasoning? Be as detailed as possible.
>
> $2B + 6R = 40$ since $B = R$, I can write
> $2B + 6B = 40$
> $8B = 40$
>
> This last equation says 8 books is equal to $40. So one book costs $5.

In this problem, students frequently interpret the letter B as a label for "books," "the number of books," "the price of the books," "the number of books times the price," and various vague combinations of any or all of the above. We have found such problems useful to stimulate class discussion on both possible ways to interpret algebraic expressions and the causes of common errors. However, it is our experience that student competence never comes easily or quickly and that it requires frequent repetition with demanding problems. A full semester of such practice is barely adequate.

OVERCOMING MISCONCEPTIONS

We conclude this section with a description of an approach suitable for use in the classroom that can be quite effective for helping students overcome misconceptions. From what we have discussed thus far, it is clear that simply telling students that their conceptual understanding of a particular mathematical topic is incorrect and then giving them an explanation is often not sufficient to extirpate the misconception. The research literature consistently indicates that misconceptions are deeply seated and not easily dislodged; in many instances, students appear to overcome a misconception only to have the same misconception resurface a short time later. This phenomenon is probably a result of the fact that when students construct learning, they become attached to the notions they have constructed (Resnick 1983). Therefore students must actively participate in the process of overcoming their misconceptions. The approach we are about to discuss attempts to elicit conflict in the students' minds deriving from an inconsistency within the misconception. The student is thereby forced to participate actively in resolving the conflict by supplanting the misconception with appropriate conceptual understanding.

We illustrate this approach within the context of the "students and professors" problem. In this problem, the conflict we wish to elicit can be achieved through a three-step process.

Qualitative understanding. The first step consists of probing for *qualitative understanding.* In this instance, students are asked whether there are more students or more professors.

Quantitative understanding. The second step consists of probing for *quantitative understanding.* For the example at hand, we would ask a question like, "Suppose there are 100 professors at this university. How many students would there be?" As stated earlier, our experience has been that the difficulties experienced by students seldom stem from an inappropriate understanding of the problem statement. Therefore, in the vast majority of cases, students will get through the first two steps without any difficulty.

Conceptual understanding. Step three consists of probing for *conceptual understanding* by asking all students in the class to write an equation that represents the relationship expressed in the problem statement. This is the range of incorrect answers that can be expected: $6S = P, 6S/P, 6S + P = T, 6S = 6P$. We have already discussed the thought processes leading students to write $6S = P$; the reasoning behind the other answers is discussed in Mestre, Gerace, and Lochhead (1982).

We now illustrate how to elicit conflict for the $6S = P$ case. Recall that we previously guided the classroom discussion so that the students agreed that there were more students than professors and further that when $P = 100$, then $S = 600$; we now ask those students who wrote $6S = P$ to check their equation by substituting $S = 600$ and see what they obtain for P. We can expect one of two possible responses to this question. Some will substitute $S = 600$ into $6S = P$ appropriately and obtain $P = 3600$, and some will ignore the equation and answer the question as in step two, namely, state that if there are 600 students, there are 100 professors. In the former instance, it is easy to elicit conflict by pointing out the contradiction in their statements, namely that earlier (in step one) they stated that there were more students than professors and that (in step two) when $S = 600, P = 100$, so how can their equation say that there are more professors than students? In the latter instance, students are demonstrating that they are not substituting appropriately into the equation they wrote but rather cuing on the problem statement and their answers in steps one and two. In this event, we actually substitute $S = 600$ into $6S = P$ and show that P results in 3600, and again we have succeeded in eliciting conflict.

In practice, this approach resembles a Socratic dialogue, since the teacher seldom tells the student what the correct answer is but simply asks probing questions that attempt to elicit a contradiction resulting from the student's misconception. The student is then guided through additional probing questions, toward reaching a resolution to the contradiction. The goal is not necessarily to have the students write the appropriate equation but to have them grapple with, and dislodge, their misconceptions so these misconceptions will not resurface at some future time.

Another virtue of this approach is that the classroom becomes a forum for some heated discussions, since it is unlikely that the entire class will agree on what the correct answer is. The instructor's job is to take on the role of moderator and let the different factions of the class argue their points of view. These types of classroom discussions are excellent not only for airing the different misconceptions that students may have but also for helping the students resolve their misconceptions though peer interaction. However, since on many problems the vast majority of students may favor an incorrect answer, the instructor must be able to ask questions that will keep the discussion alive long enough for reason to prevail. The danger is that if the discussion is cut off by an authoritative answer from the teacher, then students may generate the following rule (expressed by several of our own students): "First you write down what makes sense, then you write the reverse of that."

CONCLUDING REMARKS

It is well known that word problems have traditionally been the nemesis of many mathematics students. The translation process from words to algebra is perhaps the most difficult step in solving word problems. We have presented findings from numerous studies in cognitive research that consistently indicate that students, from beginning algebra students to college students majoring in technical fields, possess some fundamental misconceptions regarding the role and meaning of variables in translating some fairly simple two-variable word problems into equations. Now that these types of difficulties are better understood, we are hopeful that a greater effort will be devoted to diagnosing and treating these misconceptions early enough for them not to inhibit the learning of higher mathematics. It appears to us that it would be profitable for teachers to integrate into their instructional techniques an attempt both to diagnose the misconceptions that their students possess and to treat these misconceptions. Only then would we be assured that students would move on to higher-level mathematics with as little adverse baggage as possible.

REFERENCES

Clement, John. "Algebra Word Problem Solutions: Thought Processes Underlying a Common Misconception." *Journal for Research in Mathematics Education* 13 (January 1982): 16–30.

Clement, John, Jack Lochhead, and George S. Monk. "Translation Difficulties in Learning Mathematics." *American Mathematical Monthly* 88 (April 1981): 286–90.

Lochhead, Jack, Bat-Sheva Eylon, Hiroshi Ikeda, and Nand Kishor. "Representation of Mathematical Relationships in Four Countries." Paper presented at Annual Meeting of the Mathematical Association of America, 12 January 1985, Anaheim, Calif.

Mestre, José P., and William J. Gerace. "The Interplay of Linguistic Factors in Mathematical Translation Tasks." *Focus on Learning Problems in Mathematics* 8 (Winter 1986): 59–72.

Mestre, José P., William J. Gerace, and Jack Lochhead. "The Interdependence of Language and Translational Math Skills among Bilingual Hispanic Engineering Students." *Journal of Research in Science Teaching* 19 (1982): 399–410.

Mestre, José P., and Jack Lochhead. "The Variable-Reversal Error among Five Cultural Groups." In *Proceedings of the Fifth Annual Meeting of the North American Chapter of the International Group for the Psychology of Mathematics Education,* edited by J. Bergeron and Nicolas Herscovics, pp. 180–88. Montreal: International Group for the Psychology of Mathematics Education, 1983

Narode, Ronald, and Jack Lochhead. "What Do You Think?" *Impact on Instructional Improvement* 19 (Spring 1985): 56–62.

Resnick, Lauren B. "Mathematics and Science Learning: A New Conception." *Science,* 29 April 1983, pp. 477–78.

Rosnick, Peter. *Student Conceptions of Semantically Laden Letters in Algebra.* Technical Report, Cognitive Development Project. Amherst, Mass.: University of Massachusetts, 1982.

———. "Some Misconceptions Concerning the Concept of Variable." *Mathematics Teacher* 74 (September 1981): 418–20.

Rosnick, Peter, and John Clement. "Learning without Understanding: The Effect of Tutoring Strategies on Algebra Misconceptions." *Journal of Mathematical Behavior* 3 (1980): 3–27.

Soloway, Elliot, Jack Lochhead, and John Clement. "Does Computer Programming Enhance Problem Solving Ability? Some Positive Evidence on Algebra Word Problems." In *Computer Literacy Issues and Directions for 1985,* edited by Robert J. Seidel, Ronald E. Anderson, and Beverly Hunter. New York: Academic Press, 1982.

Whimbey, Arthur, and Jack Lochhead. *Developing Mathematical Skills.* New York: McGraw-Hill, 1981.

CAN YOUR ALGEBRA CLASS SOLVE THIS?

Problem 13. Find the ordered pair of numbers (x,y) for which

$$123x + 321y = 345$$

and

$$321x + 123y = 543.$$

Solution on page 248

CAN YOUR ALGEBRA CLASS SOLVE THIS?

Problem 14. Let r be the result of doubling both the base and the exponent of a^b, $b \neq 0$. If r equals the product of a^b by x^b, what is the value of x?

Solution on page 248

14

Developing Algebraic Representation Using Diagrams

Martin A. Simon
Virginia C. Stimpson

S TUDENTS often view algebra as an abstract set of operations with few ties to the real world. Although they may be able to repeat certain patterns of algebraic manipulations, many lack the understanding of algebraic concepts that are necessary to apply algebra in a wide range of problem situations (Carpenter et al. 1981). One way to foster greater understanding of these concepts is through the use of diagrams.

The following problem was used with a group of in-service teachers.

> A class was 3/5 girls. If the number of boys were doubled and 6 girls were added, there would then be an equal number of boys and girls. How many students were in the class at the outset?

The teachers were asked to find several ways to solve the problem *without* using algebra. Most of their solutions were limited to variations of guess and check. Then Anet offered her diagram solution:

I drew the large rectangle to represent the number of students in the class at the outset. We don't know how many are in the class. By dividing the class into 5 equal parts, I was able to show the 3/5 that represented the number of girls and the 2/5 that represented the number of boys.

Then I doubled the number of boys (2 more boxes) and added 6 more girls (represented by the 6 Gs).

Since the boxes are of equal size and the total number of boys and girls are equal, there should be four boxes of girls to match the four boxes of boys. The 6 additional girls

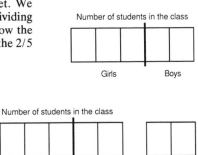

136

must be the same as one box. We know that there were 5 boxes at the outset and that each box represents 6 students, so originally there must have been 30 students.

As Anet presented this solution, "oohs" and "ahs" emerged spontaneously from her classmates. They seemed to be impressed by the clear, simple, and direct approach afforded by Anet's picture.

Peter: What Anet did is really algebra.
Teacher: Did Anet subvert the task to solve the problem without using algebra?
Lisa: No [thoughtful pause]. She proved algebra.

Lisa did not use the word "proved" in its strict mathematical sense. What she and others were expressing was that Anet had used several of algebra's basic concepts. Let us look at these concepts more closely.

1. Working with unknowns. Anet used the large rectangle to represent the unknown quantity in the problem, the number of students in the class at the outset. This permitted her to operate on the unknown quantity (3/5 of the rectangle represented the girls) using a number of manipulations that correspond to algebraic operations. The experience of manipulating the unknown quantity as if it were a known quantity is important. ("This box is the number of students in the class, so if I cut it into five equal parts, three of those will be the girls.")

2. Understanding equivalence. Researchers have documented the difficulties that the concept of equivalence poses for students (Herscovics and Kieran 1980). When Anet determined that the extra rectangle of boys was equivalent to the six additional girls, she did so by visually comparing two equal quantities. She could *see* the correspondence of the quantities of girls and the quantities of boys. The diagram thus offers students a visual experience with equivalence that is likely to be within their conceptual grasp. This experience is in sharp contrast to that of students who try to memorize what one is *allowed* to do to both sides of an equation. (Learning mathematics through imitation and memorization can contribute to a notion that mathematics is a series of magical operations.)

Students often have difficulties when they attempt to put real-world problem situations into algebraic symbols. They may be facile with the manipulation of equations but still unable to write appropriate equations from the problem statement.

For example, in research interviews conducted at the University of Massachusetts (Simon 1986), precalculus students were unable to translate information from a familiar situation (the relationship of two people's ages) to an algebraic expression. The following problem was used in the study:

> Barbara is 8 years old and Ms. Brown is 38 years old. Their birthdays are on the same day. When will Ms. Brown be three times as old as Barbara?

Though the students were asked to solve the problem using algebraic equations, they were frequently unable to do so. These same students were generally able to solve the problem by informal, arithmetic methods such as an organized guess and check. Many of the incorrect algebraic solutions were similar to the following:

$$x = \text{Ms. Brown's age}$$
$$y = \text{Barb's age}$$
$$x = 3y$$

Then they attempted to make one or both of the following substitutions: replace x by 38 or y by 8. When they obtained a result of $x = 24$, $y = 12\ 2/3$, or $38 = 24$, they realized that their algebraic solutions had not worked; yet they were unable to correct them.

An analysis of the work of these students on this and similar problems reveals two major weaknesses:

1. The students do not sufficiently understand algebraic problem solving. The fact that they set up an equation with "unknowns" and then replace these unknowns with quantities that are given in the problem shows a lack of understanding of the use of letters to represent unknown quantities. Their attempts to solve the problem with one equation containing two unknowns shows a lack of understanding of how algebra combines mathematical relationships to arrive at a particular solution.

2. The students were unable to construct an algebraic model for a familiar, real-world phenomenon: age. They apparently do not see algebra, or mathematics in general, as a medium for modeling the world. They see it instead as a procedure in which one mechanically translates the words in the problem (". . . When will Ms. Brown be three times as old as Barbara?") into corresponding algebraic symbols.

These precalculus students were doing mathematics through imitation rather than through understanding (Rosnick and Clement 1980). Instead of using what they knew about age, they were relying on a method of solution that had probably proved successful for many of the exercises they had encountered in algebra classes. In this particular problem, the translation of key words is not sufficient to provide a solution. It is necessary to focus on either the increment to the two ages (the number of years until Ms. Brown would be three times older than Barbara) or the constant difference between their ages (30 years). However, neither of these quantities is mentioned specifically in the problem. Therefore, neither was translated by the students into algebraic symbols.

The diagram solution. Diagrams can serve as a bridge between students' concrete understanding of a problem and the abstractions of algebra. When students are encouraged to illustrate the information given in the problem, they have the opportunity to use a representational system that is more familiar to them and therefore closer to the situation that they want to model. The process of drawing a diagram causes students to focus on the relevant relationships in the problem. This more concrete representation can form a solid foundation on which to build facility with algebraic representation.

Students who are asked to solve algebraic problems with diagrams *before* they learn to manipulate algebraic symbols experience algebraic word problems first as nonroutine problems rather than as exercises. Their efforts focus on developing a representation of the problem. Then, when they learn algebraic representation, they can see algebra as a more efficient and powerful method for solving a class of problems with which they are already familiar.

Initiating the drawing of diagrams in the classroom. Because many students have achieved success in mathematics classes by routinely memorizing algorithms, asking them to use diagrams may meet with resistance. Students may feel threatened by the added demands of this approach, which requires that they become more engaged intellectually. Often students have to see the limitations of a memorized approach or the power of diagrams before they are willing to work with diagrams.

Students commonly enter prealgebra and algebra classes with a less than thorough understanding of the concepts of fractions and ratio and proportion, concepts on which success in algebra depends. In addition, most students have little or no previous experience with diagrams. Time must be set aside to build the conceptual foundation for algebra and to develop the diagramming abilities of the students. It is often a good idea for the initial experiences with drawing diagrams to focus on whole-number (multistep) problems, such as this:

> Mark buys 6 shirts that cost $12 each. For every 2 shirts he buys, he gets $1 off. How much does he spend on the shirts?

When students are proficient at drawing diagrams for problems involving whole numbers, they can use this tool to draw diagrams for problems involving fractions, such as this:

> Inez has 3/4 of a gallon of ice cream. She gives Elena 2/3 of what she has. What part of a gallon of ice cream does she have left?

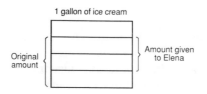

Besides developing conceptual understanding, this work fosters the realization that mathematics is meaningful. Students can develop the sense of using diagrams to model the world, and they can break out of the limited view of problem solving that is associated with applying algorithms to routine exercises (Simon 1985).

Once students have developed competence in using diagrams to solve algebraic problems, they need experiences that *link* the diagrammatic representation with the algebraic symbolization. Without this link, the benefits of diagramming often are not fully realized. Burns (1979), describing the use of concrete experience as a basis for abstract concepts, has termed this often-overlooked, intermediate step the "connecting level."

One way to provide this connecting level is to have students develop parallel representations for problems. Students first draw a part of the diagram and then record that step using algebraic symbols that correspond to the diagrammatic representation. In this way they develop a clear sense that the algebraic expressions represent, in a more abstract form, that which was represented initially by the diagram. This is demonstrated by the following problem:

> The sum of the number of books Jack and Jill have is 20. If Jill lost 3 of her books and Jack doubled the number he has, they would then have a total of 30 books. How many books does each have?

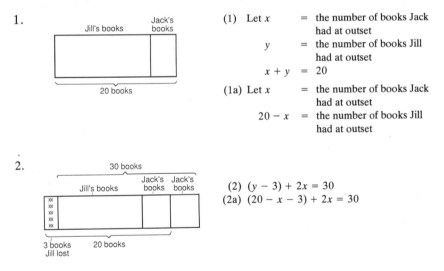

1.

Jill's books   Jack's books

20 books

(1) Let x = the number of books Jack had at outset

y = the number of books Jill had at outset

$x + y$ = 20

(1a) Let x = the number of books Jack had at outset

$20 - x$ = the number of books Jill had at outset

2.

30 books

Jill's books   Jack's books   Jack's books

3 books Jill lost   20 books

(2) $(y - 3) + 2x = 30$

(2a) $(20 - x - 3) + 2x = 30$

When a foundation for algebraic concepts has been developed and connected with algebraic symbolization, students are then ready to practice the solution of equations. Instead of viewing algebra as a study of routinized manipulations and learned algorithms, they can view the subject as meaningful and the operations as related to the more concrete diagrams to which they can turn when confusions arise with the abstractions.

The teacher and diagrams. Unfortunately, algebra and prealgebra teachers are not much more familiar with diagram solutions than their students are. They must challenge themselves to learn to solve problems diagrammatically if they wish to use this approach effectively in the classroom. Readers may wish to begin that process by returning to the problem concerning the ages of Ms. Brown and Barbara and attempting to draw a diagram to solve it. Two diagram solutions to that problem follow. Can you identify algebraic solutions that parallel the work with the diagram?

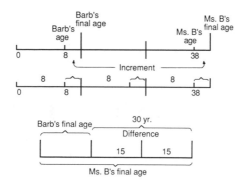

REFERENCES

Burns, Marilyn. "The Math Connection Is Yours to Make." *Learning* 7 (January 1979): 69–71.

Carpenter, Thomas P., Mary Kay Corbitt, Henry S. Kepner, Jr., Mary Montgomery Lindquist, and Robert E. Reys. *Results from the Second Mathematics Assessment of the National Assessment of Educational Progress.* Reston, Va.: National Council of Teachers of Mathematics, 1981.

Herscovics, Nicolas, and Carolyn Kieran. "Constructing Meaning for the Concept of Equation." *Mathematics Teacher* 73 (November 1980): 572–80.

Rosnick, Peter, and John Clement. "Learning without Understanding: The Effect of Tutoring Strategies on Algebra Misconceptions." *Journal of Mathematical Behavior* 3 (Fall 1980): 3–27.

Simon, Martin A. "Diagram Drawing: Effect on the Conceptual Focus of Novice Problem Solvers." In *Proceedings of the Seventh Annual Meeting of the North American Chapter of the International Group for the Psychology of Mathematics Education* (PME-NA), edited by Suzanne K. Damarin and Marilyn Shelton, pp. 269–73. Columbus, Ohio: PME-NA, 1985.

———. "An External Control Study of Diagram Drawing Skills for the Solution of Algebra Word Problems by Novice Problem Solvers." Doctoral diss., University of Massachusetts, Department of Education and Cognitive Processes Research Group, 1986.

15

Technology and Algebra

John W. McConnell

WHEN I took freshman algebra in high school more than a quarter century ago, I used William Hart's *Algebra,* the same text that my father had used a quarter century before. My oldest daughter took algebra two years earlier in school than either of us. She studied properties of real numbers, inequalities, and more coordinate graphing than I did. She was expected to have some entering knowledge of integers. But nevertheless, the topics she studied reflected the relative immutability of the subject we know as algebra. She spent considerable time on the solution of linear equations, factoring, rational expressions and equations, exponents, and polynomials. Her word problems included boats traveling upstream and downstream, parents *x* years older than their children, and whole numbers whose digits were reversed. The great emphasis in the text, the homework, and the tests was moving *x*'s and *y*'s around in a correct fashion. So even though her textbook was one that qualified as a "new math" text, the course was recognizable by her father and grandfather.

My youngest son will graduate from high school in the class of 2000. It is my hope, and my prediction, that the algebra he encounters will be very different from that of his predecessors. Technology will be a major influence, directly and indirectly, on this change, even though that belief is not held by many educators.

Today computers are well established in American schools. Data from the

142

1985 survey, "Instructional Uses of School Computers" (Becker 1986), showed that almost all secondary schools have begun to use computers in their instructional programs. The proportion of secondary schools with fifteen or more computers rose from about 10 percent to 56 percent in the two years between the 1983 and 1985 surveys. Concerning access to computers, Becker reports the median ratio of students to computers in U.S. high schools to be about 31 to 1.

We in mathematics were among the first in education to use computers. Most schools turned to a mathematics teacher when they initiated their first computer-based laboratories and programs. Yet the mathematics curriculum remains relatively untouched by technology. In practice, the algebra student rarely encounters a computer.

Teachers give a variety of reasons for not using technology. "Computers and calculators aren't on the SAT and ACT." "If you are going to study mathematics in later courses you need to know algebra well." "There aren't any materials out there that use computers." "We need to wait to see what happens as new computers and textbooks come out." "We don't have enough computers."

The last reason may be partially true. Becker has estimated that a ratio of twelve students for each machine is necessary for students to have as much as thirty minutes of computer use each day. Of course, that use is not limited to mathematics. To incorporate computers into all subjects will require significant and continuing capital expenditures. School boards are reluctant to do this.

The dilemma is clear. To get more machines, we must prove that computers are effective. To make computers effective, we must have the machines available to students. This is not a chicken/egg situation, however. The inexorable forces of change outside of education will produce changes in texts and teaching. But before one speculates on the impact of high-powered calculators and computers on algebra content, one should examine the effect of the calculator, a less imposing machine, on mathematics courses.

LESSONS FROM THE CALCULATOR

Several lessons can be learned from the study of how numeric calculators have affected mathematics. Here are the discouraging words first. Even though four-function calculators are found in every nook and cranny of the United States, many students are still subjected to topics that assume a world with no calculators, and they are forbidden to use the calculators in their classrooms.

Some school districts believe that if calculators are required, richer students will have access to better technology than poorer students. However, the cost of a *scientific* calculator is now less than the textbook used in most high school courses and, in fact, is little more than the cost of the paper a typical algebra student will consume during a year. So equity is not an issue that should deter schools from incorporating these devices in the classroom.

One curriculum project that demonstrates the capacity of technology to affect content is the University of Chicago School Mathematics Project (UCSMP). From the very first course in grade 7, *Transition Mathematics,* students are required to have a scientific calculator. This requirement creates some significant changes in content:

- *Transition Mathematics* is a third each of algebra, applied arithmetic, and geometry. The calculator enables the study of important applications.

- *Algebra* (grade 8) includes exponential growth, a topic based on the investigation of compound interest by means of the calculator.

- *Advanced Algebra* (grade 10) uses numerical patterns from the calculator to develop properties and graphs of functions.

In all UCSMP courses, *real* problems are emphasized. The calculator removes the constraints on numbers and operations that are obligatory in noncalculator courses.

All three courses develop algorithms in a calculator language as well as an algebraic language. Operations and steps are spelled out as keystrokes. Students do need explicit instruction in converting a formula into the steps of a computation. Their understanding of the variables, constants, and operations appears to be enhanced by reading and writing the steps of the calculation. Standardized testing of students studying *Transition Mathematics* showed that they maintain arithmetic competency sans calculator and are superior at solving problems *with* calculator (Hedges et al., 1986).

What we can learn from calculators is both a warning and a challenge. We see that the content of our courses does not change as fast as society embraces the technology. But when as simple a device as a calculator is assumed to be available to every child, then we can change content in ways that give effective new sequences and higher student performance.

CHANGES IN CONTENT

What changes should we expect to see in the content of algebra? Fey and Heid (1984), Fey and Good (1985), and Coxford (1985) have presented overviews of significant changes that technology should bring to mathematics—algebra in particular. These changes include a diminished emphasis on the syntactical manipulations of algebra, such as factoring, solving complicated rational expressions and equations, the analytic solution of polynomial equations, and the simplification of radical expressions.

One can expect computer and calculator graphing tools to elevate graphing to a primary position in the algebra curriculum. Fey and Good suggest that functions are central to algebra. Technology permits algebra students to answer significant questions about functions. Some of these include, What are the zeros of the function? When does the function attain a certain value? When do two functions intersect? A computer, or a calculator that graphs functions, enables students to answer these questions with a push of a button. They don't have to be put off until the senior year course or later. In *Transition to College Mathematics,* for example, Demana and Leitzel (1984) show how the numerical and graphical approach to algebra is accessible to students who intend to go to college but who have serious deficiencies in mathematics. Their text points the way to new approaches in preceding algebra courses.

The beginning algebra student can answer critical questions about functions if the content of algebra is changed in significant ways. First, students must have extensive experience with graphing in the coordinate plane. Their entry into the world of variables should be in a dynamic two-dimensional setting as opposed to a static, equation-solving one. Second, they must have more experience in translating problem situations into mathematics. The algebra course of the future will have to emphasize algebraic semantics: how to give x's and y's meaning in the context of a problem.

A rich source of problems that connect real-life situations to mathematics is found in the topics now taught as data analysis and statistics. In order for statistics to lead students to habits of modeling situations with mathematics, it must be taught early, and it must be taught in the context of the relevant mathematics. The UCSMP *Algebra* (1986) uses statistics from the very first chapter to teach and encourage the study of algebra. For example, frequency diagrams illustrate the uses of variables to describe sets of data. Scattergrams give students experience in describing patterns. The patterns are turned into formulas using the algebra of lines.

Applications of computers have moved from being task specific to general. Two kinds of programs that have become more complex, more general, and more powerful are spreadsheets and data base programs. Both exploit the dynamic nature of functions and transformations.

Spreadsheets

First, consider the spreadsheet. This programming metaphor emerged from the needs of accountants. Yet it is a mathematical tool of great power. The best-selling spreadsheet programs recognize that the user may not be able to talk about a mathematical rule in terms of x's and y's, or even with a cell name convention, like A2 or R5. They enable the user to indicate values on the spreadsheet as part of formulas. The thinking "Add this number to that one and put the answer here" can be communicated to the software by pointing to the cells using a mouse or cursor keys. When the computation is done, the user can ask the program to repeat the operation for other cells in a pattern. The program handles the conceptualization of the problem in abstract symbols. The user can solve a specific case and have the program do the generalizing. This power should give an algebra teacher pause. What do we really have to teach about variables and formulas for a person to use a spreadsheet effectively? The converse question is just as intriguing: How can we use spreadsheets to teach algebra effectively? (See Maxim and Verhey, chap. 22 of this yearbook, for suggestions.)

Data Base Programs

Another computer application with implications for mathematics is the data base. The organization of bits and pieces of data so that information can be generated through counts, sorts, lists, and links requires a spatial perception of how the data can be arranged and connected. The visualization of the data requires an understanding of function. The objects of a data base are not always real numbers, but the fundamental concepts of algebra and geometry are clearly recognizable in the basic operations on data (Liu 1985). If we accept mathematics as a subject that will prepare our students for the future, we cannot avoid the mathematics implicit in the construction and use of data bases.

There are two implications for algebra. First, we must work with functions and transformations that are not bound to the real numbers. If our students are going to use the full power of algebra, then they must encounter the mathematizing of objects other than real numbers and integers. The second implication is related. Students must learn some Boolean algebra. To use a computer to search a modern library for information, a student needs to know the mathematical meaning of *and, or, not,* and *if-then*. Boolean algebra is clearly basic to the use of technology in an information age and may be more important for our students than completing the square or solving a quadratic equation.

Changes in curriculum content will not come quickly or easily. However, we can expect that schools will continue to invest in computers for reasons not connected to the algebra curriculum. We can piggyback our needs for equipment and resources on these trends.

CHANGES IN PEDAGOGY

Computers will ultimately bring about changes in the teaching of mathematics. Three changes will occur to put more equipment in our schools and therefore produce indirect pressure on mathematics to accommodate the new technology: (1) computer-based testing, (2) the need of school districts to supplement teaching staff with computers, and (3) the increasing use of computers for writing.

Computer-based Testing

The first of these changes is computer-based testing. The deficiencies of multiple-choice tests are due to our relatively primitive technologies for measuring and scoring a large number of students on multiple objectives. A computer-based testing program will assess a student more quickly with more precision and offer more diagnostic information. The computer has the power to test the quality of the response, not just whether the answer is right or wrong, and branch the student to appropriate questions that will give more accurate information about his or her performance. So, a prediction: When testing is computer based, the public will become more interested in financing a school's investment in computers for the mathematics class.

A Supplement for Teaching Staff

A second change that will cause computers to influence teaching comes from less heartening realities. A lack of qualified mathematics teachers will make computer-based instruction a necessity for some school districts. The schools in Shakopee, Minnesota, have a large-scale computer instruction component. In algebra, students alternate class days between teacher and computer. The teacher can therefore handle twice as many algebra students.

The use of computers for instruction offers some advantages to teacher and student. Record keeping is more extensive and more complete. Diagnostic evaluation is immediate. Students can get remedial help on demand and at their own pace. The computer program can implement an effective learning theory. The disadvantage of many computer-based education systems is that they are very expensive to produce, even when they emphasize skills at the lowest cognitive levels. They enshrine the existing curriculum and are very resistant to change.

Computers for Writing

The third change that presages an increase in the number of computers in schools has to do with writing. English departments across the country are discovering that these mathematical machines are really writing tools. Students who use them tend to revise their work with more zeal and detail and hand in more readable papers.

So, to summarize, computer-based testing, the need of school districts to supplement teaching staff with computers, and the increasing use of computers for writing will put more equipment in our schools and therefore produce indirect pressure on mathematics to accommodate new technology.

SUMMARY

The adoption of technology in algebra has been a slow process. Changes in schools and society make it reasonable to suppose that assumptions retarding the adoption of technology will disappear. Then algebra will change profoundly. An algebra that exploits computers and calculators will be dynamic, graphical, and function oriented. An algebra that acknowledges the broad and emerging uses of technology will emphasize the mathematizing of a rich variety of applications. Students will have more tools available to use to solve mathematical problems.

Over twenty years ago, W. W. Sawyer (1966, p. 14) wrote:

Automation enables a machine to replace any human activity, physical or mental, that is capable of being reduced to a routine. Most present human activities are capable of such reduction, and much education is concerned with imparting routines; such education is clearly becoming obsolete. Automation will tend to concentrate human employment into occupations that call for specifically human attributes—originality, insight, judgement, initiative, understanding.

This is an appropriate time to renew and renovate our most traditional of mathematics—algebra. We have computers and calculators available in our schools to begin the process of changing, deleting, expanding, and resequencing mathematics. We must be aware that many current trends in testing, evaluation, and automation of instruction will press toward the less important and more routine aspects of our subject. Technology gives us a challenge—to move our subject away from the routine and toward the content and processes that will give the class of 2000 originality, insight, judgment, initiative, and understanding.

REFERENCES

Becker, Henry Jay. "Instructional Uses of School Computers." *Center for Social Organization of Schools* (No. 1, June 1986), Johns Hopkins University.

Coxford, Arthur. "School Algebra: What Is Still Fundamental and What Is Not?" In *The Secondary School Mathematics Curriculum,* 1985 Yearbook of the National Council of Teachers of Mathematics, edited by Christian R. Hirsch, pp. 53–64. Reston, Va.: The Council, 1985.

Demana, Franklin D., and Joan R. Leitzel. *Transition to College Mathematics.* Reading, Mass.: Addison-Wesley Publishing Co., 1984.

Fey, James T., and Richard A. Good. "Rethinking the Sequence and Priorities of High School Mathematics Curricula." In *The Secondary School Mathematics Curriculum,* 1985 Yearbook of the National Council of Teachers of Mathematics, edited by Christian R. Hirsch, pp. 43–52.

Fey, James, and M. Kathleen Heid. "Imperatives and Possibilities for New Curricula in Secondary School Mathematics." In *Computers in Mathematics Education,* 1984 Yearbook of the National Council of Teachers of Mathematics, edited by Viggo P. Hansen, pp. 20–29. Reston, Va.: The Council, 1984.

Hedges, Larry V., Susan S. Stodolsky, Sandra Mathison, and Penelope V. Flores. "Transition Mathematics Field Study." Chicago, Ill.: University of Chicago School Mathematics Project, December 1986.

Liu, C. L. *Elements of Discrete Mathematics.* 2d ed. New York: McGraw-Hill, 1985.

Sawyer, W. W. *A Path to Modern Mathematics.* Baltimore: Penguin Books, 1966.

University of Chicago School Mathematics Project texts:
Usiskin, Zalman, et al. *Transition Mathematics,* 1986.
McConnell, John W., et al. *Algebra,* formative edition, 1986.
Senk, Sharon, et al. *Advanced Algebra,* pilot edition, 1986.
Rubinstein, Rheta, and James Schultz. *Functions and Statistics with Computers,* pilot edition, 1986.

16

Computer Software for Algebra: What Should It Be?

Harley Flanders

C OMPUTER software used for mathematics instruction should be designed specifically for this purpose. Software designed primarily to produce graphs, to do numerical calculations, or to simplify algebraic expressions is inappropriate. Many of the graphical, numerical, and symbolic algebra production software packages are marvelous tools for getting results, but they are no more substitutes for instructional software than a table of trigonometric identities is a substitute for a well-written textbook chapter on trigonometric identities.

SOFTWARE STANDARDS

Teachers insist that their textbooks meet rather severe standards; they should insist on equal standards for instructional software. These are my five basic standards:

1. No knowledge of the computer should be required to use the software other than the basics: how to insert a disk, turn it on, and type in a few letters. No one should have to read a manual to get started. Certainly no computer programming whatever should be required.

2. The software should be truly interactive, with emphasis on active. Software should neither spoon-feed nor do show-and-tell demonstrations. It should require user-generated input to produce any output, thereby requiring users constantly to think about what they are doing and to be correspondingly gratified by the results of their input. Users should always feel they are part of the process. (My impression is that the kind of software that generates a random linear equation and asks for its solution, responding with "no, try again," or a hint, or "that is correct," is unsuccessful.)

3. Instructional software should be robust—extremely difficult to crash

149

or to hang. Input errors should merely be noted, with reentry requested, not be punished by the program's halting or taking some other time-wasting action. When a menu offers choices, only the letters offered should be recognized by the computer.

4. The syntax for input should be as close to the way mathematics is written as possible. For instance, one should be able to key in

$.5(2x + 1)(3x + 4)$ instead of $0.5*(2*x + 1)*(3*x + 4)$.

5. Input of real numbers should allow expressions that evaluate to reals (ditto for rationals or for integers), so that the computer does the calculations. For instance, if we want the graph of $y = x\char94 3$ to go up to height 14 exactly, we should be able to enter expressions like

$14\char94(1/3)$,   cuberoot 14,   cbrt 14

for the right endpoint of the domain.

Instructional software that satisfies these standards can serve three principal purposes:

1. It can be a tool for mathematical experiments, thereby making mathematics into a laboratory science. For instance, the computer can graph a dozen rational functions in as many minutes, whereas this task would take a student hours by hand and (usually) result in inaccurate graphs.

2. It can be a tool for removing the drudgery of routine calculations. This releases time for teaching the ideas of mathematics and for teaching the setting up of word problems, probably the hardest algebra topic to teach and to learn.

3. It can be a tool for checking hand calculations quickly and accurately.

A SAMPLE PROGRAM

Based on the standards above, a program for the solution of a two-by-two linear system might go as follows. The object of the program is to teach elimination, according to rules of procedure that generalize to higher-order systems. There are so many difficulties in teaching this central topic that it is highly desirable for problems of arithmetic not to get in the way.

First is the input section, in which we are prompted for the six coefficients of the system

(1) $a_{11}x + a_{12}y = b_1$
(2) $a_{21}x + a_{22}y = b_2$.

Let us assume the coefficients are rational; each is entered like any rational expression that evaluates to a constant. We should be allowed to correct entries until the input is satisfactory. Then the equations should be displayed.

Now we should be presented with a menu something like this:

A) Multiply an equation by a nonzero rational number.

B) Transpose the equations.

C) Add a rational multiple of equation (1) to equation (2).

D) Add a rational multiple of equation (2) to equation (1).

E) Stop.

After the first round, there should be an additional choice:

F) Return to the previous system of equations.

If, for instance, (C) is chosen, the next prompt should ask for the multiplier, and any expression that evaluates to a rational number should be accepted; anything else should result in an error message and a request to reenter the rational number.

After this input, the computer should do the arithmetic, automatically cancel common multiples, display the resulting equivalent system, and present the menu. Of course, when the system has been reduced to the form

$$(1) \qquad\qquad x = b_1$$
$$(2) \qquad\qquad y = b_2$$

a termination routine should take over.

THE MICROCALC PROGRAM

A software package with the characteristics described above, covering all topics in algebra and trigonometry, is under development. A parallel package for calculus, *MicroCalc* 3.0 for PC/XT/AT computers, has been classroom tested and is currently being distributed. Releases for Apple computers will be available later. (Information on *MicroCalc* is available from MathCalcEduc, 1449 Covington Drive, Ann Arbor, MI 48103.)

CONCLUSION

The computer will make a real impact in the classroom only with software specifically designed to meet high standards for instruction. Teachers should demand the best from educational software publishers to obtain software that is truly instructional and truly interactive.

17

Programming Finite Group Structures to Learn Algebraic Concepts

Richard J. Shumway

THIS article illustrates how computer programming can help students explore fundamental algebraic concepts, discover new insights, and appreciate the need for mathematical proof. A BASIC program to test Abelian group properties follows. Other languages would be equally appropriate. A teaching/learning sequence would involve developing and testing each procedure separately with a mathematics class.

```
1    DATA 3,1,2,3,2,1,1,3,1,1
2    READ N

10   DIM A(N, N)

20   PRINT "OPERATION TABLE:"
30     FOR J = 1 TO N
40       FOR K = 1 TO N
50         READ A(J, K)
60         PRINT TAB(4*K); A(J, K);
70       NEXT K
80       PRINT
90       PRINT
100    NEXT J

200  PRINT "COMMUTATIVITY CHECK:"
210    FOR J = 1 TO N
220      FOR K = 1 TO N
230        IF A(J, K) < > A(K, J) THEN PRINT J; K; "< >"; K; J
240      NEXT K
250    NEXT J
260  PRINT "COMMUTATIVITY CHECK DONE."

300  PRINT "ASSOCIATIVITY CHECK:"
310    FOR J = 1 TO N
320      FOR K = 1 TO N
330        FOR L = 1 TO N
340          IF A(J, A(K, L)) < > A(A(J, K),L) THEN PRINT J; "("; K; L;
             ") < > ("; J; K;")"; L
```

152

```
350        NEXT L
360        NEXT K
370        NEXT J
380        PRINT "ASSOCIATIVITY CHECK DONE."

400        LET E = 0: PRINT "IDENTITY ELEMENT CHECK:"
410          FOR J = 1 TO N
420            FOR K = 1 TO N
430            IF A(J, K) < > K OR A(K, J) < > K THEN 460
440            NEXT K
450          LET E = J: PRINT E; "IS AN IDENTITY.": GO TO 470
460          NEXT J
470        PRINT "IDENTITY CHECK DONE."

500        PRINT "INVERSE CHECK:":  IF E = 0 THEN 560
510          FOR J = 1 TO N
520            FOR K = 1 TO N
530            IF A(J, K) = E AND A(K, J) = E THEN PRINT K; "IS AN INVERSE
               FOR"; J
540            NEXT K
550          NEXT J
560        PRINT "INVERSE CHECK DONE."
```

Now the fun begins. Running the program with DATA 3,1,2,3,2,1,1,3,1,1 generates the following output:

```
OPERATION TABLE:
    1    2    3

    2    1    1

    3    1    1
COMMUTATIVITY CHECK:
COMMUTATIVITY CHECK DONE.
ASSOCIATIVITY CHECK:
2(23) < > (22)3
2(33) < > (23)3
3(22) < > (32)2
3(32) < > (33)2
ASSOCIATIVITY CHECK DONE.
IDENTITY ELEMENT CHECK:
    1 IS AN IDENTITY.
IDENTITY CHECK DONE.
INVERSE CHECK:
1 IS AN INVERSE FOR 1
2 IS AN INVERSE FOR 2
3 IS AN INVERSE FOR 2
2 IS AN INVERSE FOR 3
3 IS AN INVERSE FOR 3
INVERSE CHECK DONE.
```

According to the output, this operation is commutative, there are four instances where the operation fails to be associative, 1 is the identity element, and both 2 and 3 have two different inverses! Is the program in error? Do we not have proofs that identity elements and inverses are unique?

The proof that the identity is unique assumes there are two identities, say e and i: $e = ei = i$, and so $e = i$. We ask if one cannot also prove, by a similar argument, that the inverses are unique too. If so, how can we have more than one inverse for elements 2 and 3? Students faced with this jarring example become very interested in the two proofs.

The example shows that mathematical proofs are critical. The need for the power of proof occurs twice: once during programming, where we assumed the identity was unique (so we could call it E in line 450), and again in understanding the output.

In addition to exploring operation tables to develop more "number sense" about such properties and their independence or dependence, we can ask how common it is for a randomly generated operation to have such properties. Minor modifications in the program—adding various counters (e.g., S = S + 1) and a random-number generator (LET A(J, K) = INT(N*RND(1) +1), N being the number of elements)—allow such exploration.

You may be surprised to learn that in 100 randomly generated tables for three elements, I found one associative operation, one commutative operation, and one operation with an identity element. The frequencies become much smaller with four-element sets. Operations satisfying these properties are very special and very rare. Few students, however, are aware of their rarity. Surprisingly, nature almost always follows these Abelian group patterns.

The example illustrates one way to use technology in teaching mathematics (see Shumway [1987] for more elementary examples). Computer coding, regardless of the language, requires careful, specific knowledge of the mathematical concepts being programmed. The exact definitions must be represented in the programming code. Furthermore, the computer allows one to test knowledge of the mathematical concepts in a broader universe of examples and nonexamples than has ever before been reasonably possible. The way a computer encourages symbolic generalization, the use of variables, and a generalized view of binary operations as mappings (e.g., A(J, K)) reinforces the type of thinking about mathematics we would like to encourage in students.

REFERENCES

Shumway, Richard J. *101 Ways to Learn Mathematics Using BASIC*. Englewood Cliffs, N.J.: Prentice-Hall, 1987.

18

Relating Functions to Their Graphs

James Saunders
John DeBlassio

G RAPHING functions is a fundamental activity in mathematics teaching and learning and is a topic that can be found in all levels of high school mathematics. Students and teachers have been plotting points to analyze functions ever since René Descartes discovered analytic geometry. Plotting points and creating graphs tends to be one of the favorite activities of mathematics students. Everyone can do it! Pick an x, figure out y, plot a point, pick an x, and so forth. Ask a high school student to graph $y = x^2$, and there is a better than even chance that you will see a t table: pick an x, figure out y. Students love it! However, "Every function has a graph," is, by this procedure, replaced by, "Every function has a t table." Ask a calculus student to graph $y = x^3 + 3/x$, and chances are you will see a t table.

T tables are not all bad. They illustrate some points of a function. However, the emphasis for students and teachers should be on the graph. When students try to visualize $y = x^2$, they should see a graph, not a table of ordered pairs.

How can we emphasize the function-graph relationship? The answer is simple, and the computer can help. Let the computer figure out the values for the t table. Even better, have the computer plot the points. Then the student can concentrate on what happens to a function when changes are made, like $y = x^2$, $y = x^2 + 2$, and $y = x^2 - 2$. Graphing with the help of a computer emphasizes creativity and the inherent beauty of the finished product. Students and teachers will continue to like graphing and will gain the desirable function-graph relationship.

Many examples of good graphing programs are available for all the popular microcomputers. Sometimes these programs do more than the teacher needs or even cares about. At Upper St. Clair (Pa.) High School we use a relatively simple program created for each of the three microcomputers—Apple, Atari, and IBM. The program, which we call GRAPH, was written by students Blaze Stancampiano and Eric Graeler. Some of the ways that we have found this program to be useful follow.

When one computer is available for the classroom

1. Give students approximately three minutes to sketch the graph of a function. Then show the computer result.
2. Display a computer graph and ask the students what function is represented.
3. Work on variations of one function. Ask students what happens to $y = x^2$ when $y = x^2 + 5$; then show the computer graph.
4. Display a graph. Ask where the minimums, maximums, and inflection points are. Then try to determine what happens at these points.
5. Ask students to think up a variation on common functions whose graph is unknown, such as $y = \sin(x^2) + 5 \cos(x)$. Students should make a sketch of what they think the graph will look like, then compare it to the computer-generated version.
6. Have students use the computer as a resource for a quick sketch of a function that they may want to check in the context of a larger problem.

When a group of computers is available in a computer resource center

7. Give the students lots of practice in entering functions and making sketches directly from the monitor screen.
8. Encourage students to use the graph program any time they want to check the sketch of a function.
9. Develop worksheets that will help students achieve independently the objectives considered above in a one-computer classroom situation.

Strategies such as these will help students relate functions and their graphs. A graphing program can give students and teachers increased insight into the beauty and excitement of mathematics.

The following version of GRAPH is for the Apple computer and will produce the curve shown in figure 18.1. The function is entered on line 180 as y in terms of x. To change the scale, change line 184 as follows: $s = 2(1$ block $= 5$ units) $s = 5(1$ block $= 2$ units) $s = 8(1$ block $= 1$ unit).

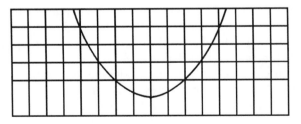

Fig. 18.1

GRAPH (for the Apple II family)

```
10 REM APPLE IIe FUNCTION PLOTTING PROGRAM
20 REM GRAPHICS MODE
30 HOME:HGR:HCOLOR = 2
40 REM PUT MARKINGS ON THE AXIS
50 FOR X = 0 TO 279 STEP 10
60 HPLOT X,0 TO X,159
70 NEXT X
80 FOR Y = 0 TO 159 STEP 10
90 HPLOT 0,Y TO 279,Y
100 NEXT Y
110 HPLOT 0,159 TO 278,159 TO 278,0
120 HCOLOR = 3
130 HPLOT 140,0 TO 140,159
139 HPLOT 141,0 TO 141,159
140 HPLOT 0,80 TO 279,80
145 E = 1
150 REM PLOT THE FUNCTION
160 FOR X = -14 TO 14 STEP .1
165 IF X = 0 THEN 200
170 X1 = 140 + X * 10
180 Y = X^2
184 S = 10
185 Y = 80 - Y * S
186 IF Y > 190 OR Y < 0 THEN E = 1:GOTO 200
187 IF E = 1 THEN HPLOT X1,Y:E = 0
190 HPLOT TO X1,Y
200 NEXT X
210 GOTO 210
```

CAN YOUR ALGEBRA CLASS SOLVE THIS?

Problem 15. The fraction $(5x - 11)/(2x^2 + x - 6)$ was obtained by adding the two fractions $A/(x + 2)$ and $B/(2x - 3)$. Find the values of A and B.

Solution on page 248

CAN YOUR ALGEBRA CLASS SOLVE THIS?

Problem 16. If the reciprocal of $x + 1$ is $x - 1$, then $x =$ _____ .

Solution on page 248

19

Computer Lessons in Algebra

Patricia Fraze

THE microcomputer is a valuable tool in the algebra class, for it allows teachers to introduce solution methods sometimes not feasible without the technology. What follows is an example of such a lesson in which a "Systematic Root Search" is performed in response to a verbally presented situation leading to a quadratic model. Initially the BASIC program is given, so no programming skills are needed. In the extensions, moderate programming skills are required, and in the additional suggested computer exercises, programming skill is assumed.

Computer Lesson: Systematic Root Search

Curriculum Topic: Quadratic Functions and Equations

Course: Algebra 1 or Algebra 2

Timing: After introductory lessons on where quadratic equations arise (e.g., projectile motion) and on the guess-and-test method of solution; before traditional methods of solution.

Equipment: One classroom monitor, or one microcomputer lab with one computer for every 2 to 3 students.

Instructions: By using a computer to systematically generate guess-and-test data, you can find the solution to a problem more quickly. When you use a computer in this way to solve an equation, you are conducting a *root search*. The solutions are often called *roots* of the equation.

Example: A toy rocket is fired straight up at 120 feet per second. The formula $h = 120t - 1 6t^2$ gives the height of the rocket above the ground t seconds after it is fired.

1. How high does the rocket go?
2. How long is the rocket in the air?

158

3. After how many seconds is the rocket 200 ft. above the starting point?

ROOT SEARCH PROGRAM	OUTPUT (FIRST = 0, LAST = 8, CHANGE = 1)	
	T	H
10 INPUT FIRST, LAST, CHANGE	0	0
20 PRINT "T", "H"	1	104
30 PRINT "____", "____"	2	176
40 FOR T = FIRST TO LAST STEP CHANGE	3	216
50 LET H = 120*T − 16*T*T	4	224
60 PRINT T, H	5	200
70 NEXT T	6	144
80 END	7	56
	8	−64

Question 1. How high does the rocket go?

To obtain a closer approximation than 224 feet, run the Root Search Program with FIRST = 3, LAST = 4, CHANGE = .1

T	H
3	216
3.1	218.24
3.2	220.16
3.3	221.76
3.4	223.04
3.5	224
3.6	224.64
3.7	224.96]←
3.8	224.96
3.9	224.64
4	224

The actual maximum height reached by the rocket seems closer to 224.96 feet.

To obtain an even closer approximation to the maximum height of the rocket, you could run the Root Search Program with CHANGE = .01. What values would you type for FIRST and LAST?

T	H
3.7	224.96
3.71	224.9744
3.72	224.9856
3.73	224.9936
3.74	224.9984
3.75	225
3.76	224.9984
3.77	224.9936
3.78	224.9856
3.79	224.9744
3.8	224.96

The maximum height is apparently 225 feet.

Question 2. How long is the rocket in the air?

There are several methods you could use to answer this question. One method is to use the Root Search Program with the values FIRST = 7, LAST = 8, CHANGE = .1, since one root of $0 = 120t - 16t^2$ is between $t = 7$ and $t = 8$. (Why?)

T	H
7	56
7.1	45.4400001
7.2	34.5600002
7.3	23.36
7.4	11.8400001
7.5	0 ⟵
7.6	−12.1599997
7.7	−24.6399996
7.8	−37.4399995
7.9	−50.5599996
8	−63.9999995

The rocket reaches the ground again after exactly 7.5 seconds.

You can arrive at this answer by another method if you remember that Galileo proved that a frictionless object projected from the ground takes as long to go up as it does to come down.

a) How long did the rocket take to go-up? (Review the output for Question 1.)

b) How long did the rocket take to go up and down?

c) How long was the rocket in the air?

Question 3. After how many seconds is the rocket 200 feet above the starting point?

One answer is given in the original table of values: after 5 seconds.

The other answer can be more accurately approximated by running the Root Search Program with FIRST = 2, LAST = 3, CHANGE = .1, since the other root (besides $t = 5$) of $200 = 120t - 16t^2$ is between $t = 2$ and $t = 3$.

T	H
2	176
2.1	181.44
2.2	186.56
2.3	191.36
2.4	195.84
2.5	200 ⟵
2.6	203.84
2.7	207.36
2.8	210.56
2.9	213.44
3	216

The rocket is 200 feet above the ground after 2.5 seconds and again after 5 seconds.

Our information about the height of the rocket is displayed in figure 19.1.

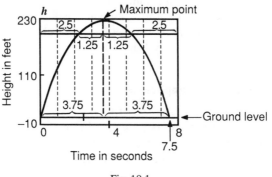

Fig. 19.1

The graph of $h = 120t - 16t^2$, for $t \geq 0$, displays two interesting characteristics:

1. The rocket reaches its maximum height after 3.75 seconds. Exactly 3.75 seconds later, the rocket hits the ground. In fact, the time at which it reaches its maximum height is the *average* of the times at which its height is 0:

$$(0 + 7.5)/2 = 3.75$$

2. The rocket reaches a height of 200 feet just 2.5 seconds after it leaves the ground and 2.5 seconds before it hits the ground. These times are equally spaced from 3.75; again:

$$(2.5 + 5)/2 = 3.75$$

Every height reached by the rocket, except 225 feet, is reached twice. If the rocket reaches a certain height at time $3.75 - t_1$ on the way up, then it reaches that height again on the way down at time $t_2 = 3.75 + t_1$. All parabolas display this characteristic, called *symmetry*. The parabola is symmetric about the line $t = 3.75$. For that reason, the line $t = 3.75$ is called the *axis of symmetry* of the parabola.

EXTENSIONS

Quadratic Maximum/Minimum Applications

1. Brian wants to build a rectangular dog run using his neighbor's fence for the longest side of the rectangular region. He has money for 60 meters of fencing. What dimensions should he use so that the dog run has as much

area as possible? (Modify the Root Search Program for use with A in terms of w.)

2. The Piedmont High School drama club charges $3 a ticket and always has a full house of 400. The club's advisor estimates that for each 25-cent increase in the ticket price, the ticket sales will decrease by 15. What ticket price would earn the most revenue for the club? (Modify the Root Search Program for use with S in terms of x, the number of 25-cent increases.)

Cubic Maximum/Minimum Applications

If you were to cut small squares from the corners of a 16 cm × 20 cm sheet of metal, you could fold the remaining metal to make a box without a top. The volume of the box depends on x, the size of the squares you cut out. If you want to maximize the volume of the box, how big should x be?

Approximating Irrational Roots of Polynomial Equations

Approximate the irrational roots of $x^3 - 3x^2 - 7x + 5 = 0$ to three decimal places. (Run the Root Search Program for integers, then for .1, .01, .001, and .0001 in the appropriate intervals.)

SUGGESTED COMPUTER EXERCISES FOR ALGEBRA 2 AND TRIGONOMETRY

Options

1. Assign one or more computer exercises along with related textbook exercises.
2. Assign small groups to write or write and run the program in the classroom.
3. Discuss the program in class; use the monitor to run.

Exercises

1. Write a program to solve $ax + b < c$ or $ax < bx + c$. (*Challenge:* Solve $|ax + b| < c$ or $> c$.)
2. Consider the SLOPE program below:

```
10   REM SLOPE OF LINE SEGMENT
20   INPUT X1, Y1, X2, Y2
30   LET M = (Y2 − Y1)/(X2 − X1)
40   PRINT "SLOPE = " M
50   END
```

   a) Modify the SLOPE program to handle the situation $X2 = X1$.
   b) Write a program to determine whether $\overleftrightarrow{AB} \parallel \overleftrightarrow{CD}$, given the coordinates of A, B, C, and D.

c) Write a program to determine whether *ABCD* is a parallelogram, given the coordinates of *A, B, C,* and *D.*

3. Write a program that uses Cramer's rule (determinants) to solve a system of two linear equations:

$$\begin{cases} Ax + By = C \\ Dx + Ey = F \end{cases}$$

Include all three possibilities: one point, a line, or the empty set.

4. Write a program that accepts the coefficients of a quadratic function, $y = Ax^2 + Bx + C$ and—

   a) prints the coordinates of the vertex;

   b) prints the equation of the axis of symmetry;

   c) prints whether the vertex is a maximum or minimum point.

5. Write a program that accepts the coordinates of three points and uses the distance formula to determine—

   a) whether a triangle is formed;

   b) if so, whether it is scalene, isosceles, or equilateral.

6. Write a program to print *n* arithmetic means between two given numbers. *(Challenge: n* geometric means)

7. Write a program to print the first *n* terms of the Fibonacci sequence.

8. Write a program to solve for the missing sides and angles of a triangle, given—

   a) *SSS*   *b*) *SAS*   *c*) *ASA* or *AAS*

9. Write a program to determine the number of triangles formed, given *SSA.*

10. Write a program to simulate *n* tosses of a coin or die or other polyhedron and to tabulate the results.

SAMPLE PROGRAM FOR COIN TOSS

(Note: RND(1) returns a random number between 0, inclusive, and 1.)

```
 10  INPUT N
 20  REM N IS NUMBER OF TOSSES
 30  T = 0: H = 0
 40  FOR J = 1 TO N
 50  R = INT(RND(1)*2) <------ R is either 0 or 1
 60  IF R = 0 THEN T = T + 1
 70  IF R = 1 THEN H = H + 1
 80  PRINT "RESULTS OF " N "TOSSES: "
 90  PRINT T "TAILS AND" H "HEADS."
100  END
```

20

Computer-calculated Roots of Polynomials

Alfinio Flores

FACTORING to find the roots of polynomials is a widely taught method. Textbooks allocate much space to it, and classroom teachers devote a lot of time to teaching it. However, this method almost never works, except for the examples in the textbook. Very few of the polynomials that appear in applications are easily factorable, yet they all are, theoretically.

The computer can be used to demonstrate how unusual factorability over the integers is. The BASIC program RANDPOLYNOM randomly generates integers as coefficients of polynomials and checks whether the resulting polynomial is factorable over the integers. The program lists the coefficients of the polynomial and tells if it is factorable.

Program RANDPOLYNOM

```
10 REM RANDPOLYNOM
20 REM GENERATES COEFFICIENTS OF POLYNOMIAL AT RANDOM
30 REM  X ^ 2 + BX + C = 0     -100 < B,C < 100
40 REM CHECKS WHETHER POLYNOMIAL IS FACTORABLE OVER THE
   INTEGERS
50 REM ***************************************************************
70 FOR N = 1 TO 100
80 LET B = INT (200 * RND(1)) - 100
90 LET C = INT (200 * RND(1)) - 100
100 LET D = B*B - 4*1*C
110 IF D < 0 THEN PRINT 1;" ";B;" ";C;" COMPLEX ";:GOTO 210
120 LET E = (-B + SQR(D))/2
130 LET F = (-B - SQR(D))/2
140 IF ABS (INT(E) - E) < 0.000001 OR ABS(INT(E)+ 1 - E) < 0.000001
   THEN LET V1 = 1
150 IF ABS (INT(F) - F) < 0.000001 OR ABS(INT(F)+ 1 - F) < 0.000001
   THEN LET V2 = 1
160 LET V = V1 *V2
170 IF V = 1 THEN LET M = M+1: PRINT 1;" ";B;" ";C," YES ",
180 IF V = 0 THEN PRINT 1 ;" ";B;" ";C,
190 LET V1 = 0 : LET V2 = 0
```

164

```
210 PRINT N;" ";M/N
220 NEXT N
230 END
```

It is surprising how few of the resulting polynomials are factorable over the integers even though they are factorable over the complex numbers. An interesting problem for students is to estimate the probability of a polynomial generated by RANDPOLYNOM being factorable over the integers.

One can find roots of a polynomial with other methods that illustrate powerful problem-solving techniques and that are compatible with technology. One such method, *successive approximations,* can be described in four steps:

1. Approximate a root.
2. Check whether your approximation is good enough.
3. Get a better approximation from the previous one.
4. Repeat steps 2 and 3.

Computers are particularly well suited for this method, since they compute rapidly and can easily repeat the same procedure many times.

THE BISECTION METHOD OF APPROXIMATION

For a polynomial P, if a and b are numbers such that $P(a) < 0$ and $P(b) > 0$, then there is a value r between a and b such that $P(r) = 0$. Each such number r is a root of the polynomial. (See fig. 20.1.)

The numbers a and b are the first guesses used to determine an interval where a root is located. To get a closer approximation for r, we can evaluate the polynomial at the midpoint of the interval. If the polynomial is zero, we have a root. If not, we know that the function changes sign in one of the halves. In the example above, the change is in the right half. In general, if $P(a)$ and $P((a + b)/2)$ have different signs, we take $(a + b)/2$ as our new right endpoint; otherwise, we take $(a + b)/2$ as our new left endpoint. (See fig. 20.2.)

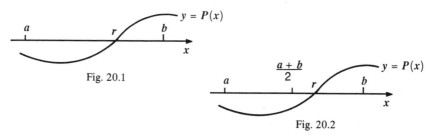

Fig. 20.1

Fig. 20.2

Now we are closer to the solution. We have the same situation as originally but with an interval of half the length. The procedure is repeated producing smaller and smaller intervals until we determine the value of the root to the desired degree of precision. This is the *bisection method* of approximation.

Student Activity 1

Use the program ROOTS1 to approximate the roots of $x^3 + 3x^2 + 4x - 11$. Run the program to get a value r close enough to a root for $P(r)$ to be less than 0.0001 in absolute value. How many times did the computer repeat the procedure to get the desired approximation?

Program ROOTS1

```
10 REM ROOTS1
20 REM BISECTION METHOD
100 REM  ***********************************************************
110 DEF FN Y(X) = X*X*X + 3*X*X + 4*X - 11
120 INPUT " INTERVAL"; A, B
130 INPUT "E "; E
140 PRINT "R", "Y(R)"
150 LET R = (A + B) / 2
160 PRINT R , FN Y(R)
170 IF ABS (FN Y(R)) < E THEN END
180 IF FN Y(R) > 0 THEN LET B = R
190 IF FN Y(R) < 0 THEN LET A = R
200 GO TO 150
210 REM  ***********************************************************
```

Suppose now that you want more precision, for example, $P(r)$ to be less than 0.0000001 in absolute value. How many more times would the procedure be repeated? The answer can be deduced. Since the length of the interval is halved each time and $2^{10} = 1024$, the procedure needs to be repeated ten times to get an interval that is 1/1000 of the last interval. So if you want three more digits of precision, the computer must repeat the procedure roughly ten more times.

Student Activity 2

Approximate the roots of these polynomials. Change line 110 of ROOTS1 to define the appropriate function.

a) $x^5 + 2x^4 - 4x^3 - 2x^2 + x + 3$

b) $x^5 + x^4 + 1.2x^3 + 0.7x^2 - x + 2$

Student Activity 3

ROOTS1 can also be used in the case where $a < b$ and $P(a) > 0$ and $P(b) < 0$. To get a root of $-x^3 + 3x^2 + 4x - 11$, give the endpoints of the interval in the reverse order, for example, $10, -10$.

If it is desired to evaluate a polynomial of any integral degree, how should the computer compute the values? This is an important question for computer scientists. It is better to use multiplication of numbers rather than exponentiation because multiplication is faster and more precise. For higher-order polynomials it is too cumbersome to write

$$x*x*x*x*x*x*x + x*x*x*x*x*x + \ldots$$

Fortunately there is a way to compute a polynomial without having to multiply so many times. For example, instead of computing

$$6*x*x*x*x + 5*x*x*x* + 4*x*x + 3*x + 2$$

(9 multiplications and 4 sums), we can tell the computer to compute

$$2 + x*(3 + x*(4 + x*(5 + x*6)))$$

(4 multiplications and 4 sums).

In general, to evaluate the polynomial

$$a_n x^n + a_{n-1} x^{n-1} + \ldots + a_2 x^2 + a_1 x + a_0,$$

calculate

$$a_0 + x*(a_1 + x*(a_2 + x*(\ldots))).$$

The number of multiplications grows like n, the degree of the polynomial, whereas with the usual method the number of multiplications grows like $n^2/2$. This is Horner's method.

This way of evaluating a polynomial is seldom taught in high school. One important aspect of computing is that usually there are several procedures to solve a problem, and some of them are more efficient, faster, or more precise than others. Most people are impressed by how much faster computers are now than they were forty years ago. However, there has been even greater progress in the numerical procedures used in computing. In the same span of time that computers became faster by a factor of 10^5, in some fields the algorithmic procedures were made faster by a factor of 10^{10} (Boggs 1981).

Student Activity 4

Write the following polynomials in the form $a_0 + x * (a_1 + x *(a_2 + x* (\ldots)))$:

  a) $4x^5 - x^4 - 4x^3 - 2x^2 + x + 3$
  b) $-0.4x^3 - 2x^2 + x + 3.14159$

It is very simple to tell the computer to evaluate a polynomial this way. First we tell the computer to take P as a_n. Then for $k = 1$ to n, instruct the computer to take P as $P*x + a_{n-k}$. Program ROOTS uses this method as a subroutine to evaluate a polynomial given its coefficients.

Program ROOTS

```
10 REM ROOTS
20 REM BISECTION METHOD POLYNOMIAL ANY DEGREE
```

```
100 REM   *********************************************************
110 REM MAIN PROGRAM
120 GOSUB 2000
130 INPUT " INTERVAL"; A , B
140 LET E = .0000001
150 PRINT "R" , "Y(R)"
160 LET R = (A + B) / 2
170 GOSUB 1000
180 IF ABS (Y) < E THEN END
190 IF Y > 0 THEN LET B = R
200 IF Y < 0 THEN LET A = R
210 GO TO 160
220 REM   *********************************************************
1000 REM SUBROUTINE TO EVALUATE
1010 LET Y = C(N + 1)
1020 FOR J = N TO 1 STEP −1
1030 LET Y = R * Y + C(J)
1040 NEXT J
1050 PRINT R , Y
1060 RETURN
2000 REM   *********************************************************
2010 REM DATA
2020 INPUT "DEGREE: "; N
2030 DIM C(N + 1)
2040 FOR J = 0 TO N
2050 PRINT "COEFFICIENT X ^ ";J,
2060 INPUT C(J + 1)
2070 NEXT J
2080 RETURN
2090 REM   *********************************************************
```

Student Activity 5

Use ROOTS to find at least one root of each of the following polynomials:

a) $x^7 + 2x^6 - 4x^5 - x^4 - 4x^3 - 2x^2 + x + 3$

b) $x^5 - 0.2x^4 - 0.4x^3 - 2x^2 + x + 3.14159$

CONCLUSION

One of the goals of mathematics teaching should be for students to learn problem-solving methods and skills that are powerful and that can be used in a wide range of problems. Factoring is of limited usefulness in real problems. Thus, approximation techniques should be taught—and factoring should be de-emphasized—as a method to obtain roots (Usiskin 1980). The bisection method of obtaining roots of polynomials has several important characteristics. It is an iterative procedure where our approximations get closer and closer to the desired value. We can get the degree of precision wanted by repeating the procedure enough times. This method uses important ideas of mathematics (the theorem of intermediate value and the concept of nested intervals). The bisection method can be used to find zeros of any continuous function. It is easy to understand and easy to remember. It is a nice way to teach educated guessing. It exemplifies a powerful idea, namely, guess and have a way to improve on your guess.

BIBLIOGRAPHY

Boggs, Paul T. "Mathematical Software: How to Sell Mathematics." In *Mathematics Tomorrow,* edited by Lynn Arthur Steen, pp. 221–29. New York: Springer-Verlag, 1981.

Fey, James T., ed. *Computing and Mathematics: The Impact on Secondary School Curricula.* Reston, Va.: National Council of Teachers of Mathematics, 1984.

Johnson, Jerry. "Algorithmics and the Mathematics Curriculum." Paper presented at the Annual Meeting of the National Council of Teachers of Mathematics, San Antonio, April 1985.

Snover, Stephen L., and Mark A. Spikell. "Generally, How Do You Solve Equations?" *Mathematics Teacher* 72 (May 1979): 326–36.

Sullivan, Jillian C. F. "Polynomial Equations Revisited." *Mathematics Teacher* 79 (December 1986): 732–37.

Usiskin, Zalman. "What Should *Not* Be in the Algebra and Geometry Curricula of Average College-bound Students?" *Mathematics Teacher* 73 (September 1980): 413–24.

CAN YOUR ALGEBRA CLASS SOLVE THIS?

Problem 17.

If $\dfrac{1}{2 - \dfrac{x}{1 - x}} = \dfrac{1}{2}$, then $x = $ _____ .

Solution on page 248

CAN YOUR ALGEBRA CLASS SOLVE THIS?

Problem 18. If $f(x) = ax^2 + bx + c$ passes through the points $(-1,12)$, $(0,5)$, and $(2,-3)$, find the value $a + b + c$.

Solution on page 248

21

Computer-generated Tables: Tools for Concept Development in Elementary Algebra

M. Kathleen Heid
Dan Kunkle

E SSENTIAL to the study of elementary algebra is the conceptual under-standing of algebraic expressions in one and two variables. Although the standard high school algebra course devotes major attention to procedural aspects of the evaluation and comparison of polynomial expressions in one and two variables, the average student commonly emerges from this experi-ence with little real ability to exploit the use of these expressions in prob-lem-solving settings. In this article we describe our use of computer-gener-ated tables to address three kinds of inadequacies in students' understanding of algebraic expressions.

The first of these weaknesses relates to the fundamental notion of change. Since algebraic expressions contain variables that take on a range of values, the expressions themselves also take on a range of values. Too often, however, students perceive the variables and variable expressions only as representations of fixed numbers. What is inherently a dynamic concept is reduced to a static one. A second weakness lies in the fact that the focus of traditional school algebra on one manipulative technique at a time leaves students with a shaky grasp of fundamental ideas relating to algebraic statements of comparison. As the course's focus moves from one technique to the next, there is little occasion to compare or contrast the meanings of the concepts of expression, equation, and inequality. As a consequence, stu-dents are often unable to distinguish between equations and the expressions that make up the equations, and they frequently fail to see relationships between equations and inequalities. Finally, with the typical dominance of manipulative techniques in elementary algebra, students come to view the real substance of algebra as the execution of fixed sequences of symbol manipulations rather than as a mathematical aid to problem formulation and interpretation. Teachers have long recognized these difficulties and, in less optimistic times, have viewed them as unavoidable occupational hazards.

USING THE COMPUTER TO GENERATE TABLES

Our experience with computer-based curriculum modules (Heid 1986) in an elementary algebra classroom (imbedded in a traditional curriculum) over a six-month period suggested that classroom use of computer-generated tables of values affords the elementary algebra teacher an unprecedented opportunity to strengthen students' understanding in each of the three areas mentioned above. The modules provided students with a computer program that automated the production of tables of values, the evaluation of expressions, and the solutions of equations and systems of equations. The classroom setting was designed to facilitate both teacher demonstration and student hands-on activity with these programs. The class used Apple IIe's, each with 48k of active memory. The particular software used for this course was a muMath[1] package limited to four functions: TABLE, TABLE2, SOLVE, and SOLVE2. The TABLE function generates a table of variable values and corresponding expression values for a given expression over a given range of the variable. For example, the command could be given to generate a table of values for the expression $x^2 - 10x + 30$ for values of x ranging from 1 to 19 in increments of 2. The result is shown in table 21.1.

TABLE 21.1

1	21
3	9
5	5
7	9
9	21
11	41
13	69
15	105
17	149
19	201

The TABLE2 function performs similarly for two expressions in one variable over the same range of variable values. SOLVE will return the solution set of a variety of equations in one variable (including all linear and quadratic equations), and SOLVE2 will return solutions to linear-linear and linear-quadratic systems of two equations in two variables. Although the SOLVE and SOLVE2 commands are unique to symbol manipulation programs like muMath, table programs are easy to construct in a variety of computer languages and are widely available on the commercial market. The promise of table programs as aids in algebra problem solving has been suggested earlier (Fey 1984), and our work with the TABLE and TABLE2 commands lends further corroboration. With computer-generated tables as tools, the curriculum could attain a clearer focus on the dynamic nature of algebra; on the meanings of equations, expressions, and solutions; and on the nonmanipulative aspects of algebra problems. The particulars of that refined focus are discussed in the sections that follow.

FOCUS ON THE DYNAMICS OF CHANGE

With computer-generated tables, algebra instruction could concentrate almost immediately on how values of algebraic expressions change. In the computer-enhanced classes, students began their study of problem situations and tables of values within the first two weeks of the course. The early use of tables allowed students to explore problem situations earlier and more freely than in the traditional curriculum. A beginning problem situation, for example, dealt with the running time of Art, an athlete:

> Art is a cross-country runner. He runs at a steady pace of 210 meters per minute. The time Art takes to finish a course is dependent only on the length (L meters) of the course.

Students produced tables of values (for $L = 1000$ to 4000 in steps of 250) for Art's running time ($L/210$ minutes) as a function of the length of the course (L meters) and answered the connected set of queries produced here:

> How much more time does it take Art to run 2750 meters than to run 1750 meters?
>
> How much more time does it take Art to run 3000 meters than to run 2000 meters?
>
> Estimate how long it would take Art to run 5000 meters.

A second problem situation dealt with the time it takes to travel from Pittsburgh to Washington, D.C.:

> Pittsburgh is 245 miles from Washington, D.C. We want to analyze the amount of time the trip will take at different average speeds (S).

Students examined a table of values for travel time ($245/S$ hours) as S (mph) ranged from 10 to 70 mph (in increments of 5). Related questions were discussed:

> How much less time would the trip take a car traveling 55 miles per hour than one traveling 50 miles per hour?
>
> How much less time would the trip take a car traveling 45 miles per hour than one traveling 40 miles per hour?

In this way, students were directed to observing related rates of change. In the running-time problem, for example, increases of 1000 meters in running distances will produce constant changes in running times of about 4.76

minutes. In the driving-time problem, however, increases of 5 miles an hour will produce different changes in driving times. Students considered these differences and discussed them in light of the form of the expression involved. They noted that an expression (like $245/S$) with the variable in the denominator did not produce the constant changes associated with expressions (like $L/210$) whose variable was in the numerator. They recognized that the form of an algebraic expression influenced its rate of change. In this fashion, the treatment of algebraic expressions emphasized the dynamics of change.

A second computer-based unit, implemented during the fifth and sixth months of the course, refined this focus. During this unit, students examined differences in successive function values and determined computational rules for changes in table values for linear and quadratic expressions. They readily extended their observations to polynomial expressions of higher degree. From the tables they produced, students determined domains over which functions appeared to be increasing or decreasing and noted rates at which function values were changing. Students then used the properties they discovered to compare simultaneous changes in tables for the value of the two expressions.

For example, students examined tables like table 21.2 and noted that expression values would probably be equal for some x-value greater than 5. As several students remarked, although the first expression had larger table values than the second, the second expression would eventually "catch up." They explained that the second expression was growing faster, whereas the first expression was growing at a constant rate! With access to computer-generated tables of values, students not only focused on the dynamic features of algebraic expressions but also learned to work with these features effectively.

TABLE 21.2

x	$7x + 9$	$x^2 + 1$
1	16	2
2	23	5
3	30	10
4	37	17
5	44	26

FOCUS ON THE RELATIONSHIPS AMONG EXPRESSIONS, EQUATIONS, AND INEQUALITIES

In the traditional approach to equation solving, the concept of "solution" is obscured by an emphasis on procedure. Unlike the traditional approach, the table approach to solving equations keeps the concept of solution in sight at all times. For the students in the computer-enriched class, an expression was represented by the array of values that could appear in a single column of a table. An equation, like $x^2 + x = 2$, focused attention on a single value in that column. For example, the student would focus on the second and fifth lines of table 21.3 to find solutions to $x^2 + x = 2$. An equation like $2x + 3 = 4x + 9$ focused attention on the row for which TABLE2 produced equal expression values. Equations were not cryptic symbols to be moved about a page according to a memorized ritual; rather, they were meaningful statements about expression values. Even in their subsequent work with traditional algebra techniques, students in the computer-enriched class seemed to have little difficulty relating the concepts of expression, equation, and solution.

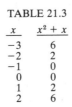

TABLE 21.3

x	$x^2 + x$
−3	6
−2	2
−1	0
0	0
1	2
2	6

The meaning of equations was further clarified through its natural extension to the meaning of inequalities. Using TABLE and TABLE2 before learning formal rules for manipulating inequalities, students could develop a perception of inequalities that was parallel to the one discussed above for equations. Inequalities were, like equations, meaningful statements about expression values (often associated with an entire range of expression values). When students were asked to solve the inequality $x^2 > x + 6$, for example, they examined a table like table 21.4. They then searched for the values of x that generated the appropriate relationship between x^2 and $x + 6$. Reasoning from what they had learned previously about linear and quadratic trends, they recognized the solution as the set of x-values less than −2 or greater than 3.

TABLE 21.4

x	x^2	$x + 6$
−5	25	1
−4	16	2
−3	9	3
−2	4	4
−1	1	5
0	0	6
1	1	7
2	4	8
3	9	9
4	16	10
5	25	11

The use of computer-generated tables highlighted the relationships among expressions, equations, and inequalities. When concepts are not treated in isolation, distinctions among them are more natural. Students learn to view equations and inequalities as conditions applied to expressions.

FOCUS ON MEANING RATHER THAN MECHANICS

Too often the treatment of word problems in elementary algebra centers on prescriptions about the technique of word-to-symbol translation for particular types of problems. The result is often a fragile understanding of underlying meaning. Using computer-generated tables permitted the computer-enriched class to shift some attention away from mechanical manipulations and toward the development of meaning. The shift was particularly salient in two facets of the curriculum: application problems and interpolation.

The computer-enriched class began a discussion of application problems virtually at the start of the course. The immediate ability to produce accurate and reliable computer-generated tables of values provided an environment for constant and in-depth attention to applications throughout the year. For a problem situation, some of the time saved by the computer generation of table values can be reallocated to extensive discussion of the formulation of symbolic expressions. For example, here is the initial discussion of the travel-time problem described earlier:

> Pittsburgh is 245 miles from Washington, D.C. Superhighways connect the cities, but the average speed a car goes depends on things like road conditions, weather, traffic, and the condition of the car. We want to analyze the amount of time the trip will take at different rates of speed.
>
> First, we must identify the quantities of interest. In this case we are interested in the average *speed* of the car as it affects travel *time*.
>
> Second, we use a variable to represent one of the quantities and find an algebraic expression for the second quantity in terms of that variable. In this example, speed and time are quantities of interest. We let speed be represented by the variable s, and we will try to find an algebraic expression for time in terms of s.
>
> To help us find an algebraic expression, it often helps to look at a few numerical examples. If we know the speed of a car traveling 245 miles, how can we find the time of travel? If the car were traveling 50 miles per hour, it would travel 100 miles in 2 hours, 150 miles in 3 hours, and so on. To find out the number of hours to travel 245 miles, we would have to find out how many 50's are in 245 (we divide 245 by 50). Since $245/50 = 4.9$, we conclude that it would take approximately 4.9 hours for a car traveling 50 mph to go from Pittsburgh to Washington. In general, the travel time (in hours) will equal 245 divided by the car's average speed (in miles per hour). Our formula is "travel time = $245/s$."

Throughout the computer modules, careful attention was paid to process (the need to identify and symbolize relevant quantities) and to meaning (concrete numerical examples were used to lead up to the formulation of the more abstract algebraic expressions).

Similar care was taken in the treatment of the interpretation of results. Problem situations invariably led to series of questions about the situations and the algebraic expressions that represented them. In the context of a single situation students could be asked to identify expression values for given variable values and to identify variable values that produced expression values meeting stated conditions. Access to tables allowed students to discuss the effect of a change in the variable value on the value of expressions containing that variable:

> If Art takes $L/210$ minutes to run L meters, by how many minutes does he increase his time for each additional 100 meters he runs?

Some of the time usually allocated to computing function values was devoted to interpreting symbolic expressions in terms of the quantities they represented:

> A videotape rental store bills its customers at the end of every month. To figure out the total bill (in dollars) for a customer, the store owner uses the algebraic expression $10 + 2.95x$, where x is the number of tapes the customer has rented that month. Explain the meaning of the numbers in the expression in words that the customer would understand.

Work with application problems consistently concentrated on building a broader understanding of the problem situation.

A second facet of elementary algebra that is transfigured by the computer-based modules is one that plays a minimal (but underrated) role in traditional algebra instruction. The concept of interpolation generates a large majority of the real-world applications of algebra. Variables defined on continuous domains commonly undergird elementary algebra applications, and interpolations become an important part of the work in formulating questions and interpreting results. A solid understanding of interpolation becomes even more important in the tables approach to algebraic expressions.

Consider, for example, the typical second-year algebra problem of solving the linear-quatratic system

$$y = x^2 - 6x + 8$$
$$y = (x - 2)/2.$$

The students in the computer-enriched first-year algebra course solved such systems using the TABLE2 command. A student could proceed by conducting an initial examination of a table like table 21.5.

Armed with the understanding that linear-quadratic systems have at most two solutions, the student would recognize that one of the solutions was

(2,0) and then would search for a feasible region for the possible second solution. Reasoning that the expression values approached each other in the vicinity of $x = 4$, the student might continue the search by producing a table like table 21.6. Without further search, the student would give the solutions (2,0) and (4.5, 1.25). As students in the computer-enriched class solved systems like these, they built on, and developed a rationale for, their intuitive understanding of interpolation.

TABLE 21.5

x	$x^2 - 6x + 8$	$(x - 2)/2$
0	8	−1
1	3	−0.5
2	0	0
3	−1	0.5
4	0	1
5	3	1.5
6	8	2
7	15	2.5
8	24	3
9	35	3.5
10	48	4

TABLE 21.6

x	$x^2 - 6x + 8$	$(x - 2)/2$
3	−1	0.5
3.5	−0.75	0.75
4	0	1
4.5	1.25	1.25
5	3	1.5

IMPLICATIONS FOR CLASSROOM PRACTICE

An important feature of our implementation of the computer-based modules was the fact that students had individual access to the table programs on a daily basis. A typical instructional strategy started with an in-depth group discussion of a problem situation and continued with the classroom demonstration of some initial analysis of the problem. Students then continued the analysis as they worked through module material at the available computer stations. Individual access to the software provided the setting for hypothesis generation and testing.

Computer programs to generate tables of values are readily available to classroom teachers. With the early introduction of such programs in the first-year algebra curriculum, teachers can engage students in the exploration of a variety of meaningful problem situations prior to subsequent work with algebraic techniques. Throughout the elementary algebra curriculum, continued work with tables of values can provide concrete numerical examples to strengthen the vision of algebra as the study of the dynamics of changes in related variables. Thoughtful use of computer-generated tables offers intriguing promise for enhancing students' understanding of algebraic expressions.

REFERENCES

Fey, James T., ed. *Computing and Mathematics: The Impact on Secondary School Curricula.* Reston, Va.: National Council of Teachers of Mathematics, 1984.

Heid, M. Kathleen. "Elementary Algebra with muMath." Unpublished manuscript, 1986.

[1] Information on muMath for the Apple IIe can be obtained from The Software House, Honolulu, Hawaii.

22

Using Spreadsheets in Algebra Instruction

Bruce R. Maxim
Roger F. Verhey

SPREADSHEET programs can be effective teaching tools for helping students experience the process of doing mathematics. A number of topics can be introduced in a meaningful way using a spreadsheet program. Two such topics, finding the roots of a polynomial using an iterative process and finding the solution of a mixture problem using an inductive process, are discussed here.

A spreadsheet is a two-dimensional array of cells. Each cell can hold a label, a value, or an expression. Expressions may contain constants or values from other cells. Spreadsheet programs provide the capability of copying expressions from one cell to any number of other cells. The cell references in these expressions can be adjusted automatically while the program handles the replication request.

As a first illustration, consider finding approximations for the roots of a polynomial equation. A common technique is to use an iterative process producing successively smaller intervals in which the value of the polynomial changes sign. This process may be continued to obtain a value with the desired accuracy. Figure 22.1 contains a spreadsheet model that may be used to solve the polynomial equation $x^3 - 2x^2 + x - 1 = 0$. One begins by using whole-number values to determine the endpoints of the change-of-sign intervals. This information is then used to determine a new starting point and a new increment. For this example, the next step might be to use 1.0 as the starting point and 0.1 as the increment.

In figure 22.1 the user enters values in cells B3 and B4. The spreadsheet program computes the values in columns C and D. Hence expressions must be entered in these columns. For example, the expression entered in cell C3 would be " +B3" (which would have 0 as its value); in C4 the expression would be "+C3+B4" (which would have 1 as its value). An appropriate representation of the polynomial then appears in cells D3 . . . D10.

	A	B	C	D
1:			INTERVAL ENDPOINT	X^3−2*(X^2)+X−1
2:				
3:	STARTING POINT	0	0	−1
4:	INCREMENT	1	1	−1
5:			2	1
6:			3	11
7:			4	35
8:			5	79
9:			6	149
10:			7	251

Fig. 22.1. Polynomial roots

Using a spreadsheet allows the student to be involved in the iterative solution process of the polynomial. Students are given the opportunity to analyze the information contained in the spreadsheet model and to continue the process as necessary. The spreadsheet program does the calculations, and the spreadsheet model organizes the newly computed information.

A second illustration of the use of a spreadsheet program deals with the solution process for a mixture problem. In this example a spreadsheet program will be used to construct a model for this problem: "How many liters of water must be added to change the concentration of salt in 10 liters of solution from 20% to 15%?"

The solution process begins with class discussion to determine the essential parts of the problem. These parts are used to develop column labels for the spreadsheet model. The labels for the columns in figure 22.2 are a typical outcome from such a discussion. The purpose of this discussion is to help students develop the ability to break a problem into its component parts.

	A	B	C	D	E	F	G	H	I
1:	LITERS		LITERS		LITERS		LITERS		% SALT
2:	SALT WATER		WATER		NEW		SALT		NEW
3:	ORIGINALLY		ADDED		MIXTURE		MIXTURE		MIXTURE
4:									
5:	10		0		10		2		20.0
6:	10		1		11		2		18.2
7:			2						
8:									
9:									
10:									
11:									
12:	10		X		10 + X		2		2/(10 + X)*100

Fig. 22.2. Mixture problem

The next step in the process is to enter the information given in the problem statement into cells under the appropriate column headings. This is done in row 5 of figure 22.2. The computation of the value in cell I5 should be considered at this point in the development of the model. The value of I5, expressed as a percent, is "$(G5/E5)*100$".

Now the class discussion should turn to the identification of the components that change as water is added to the original salt solution. This might be accomplished by placing a value in C6 and asking what values need to be entered in the remaining cells in row 6. For example, the entry for E6 should be an expression that computes the sum of C6 and A6. Students may need some help in understanding why the values entered in cells A6 and G6 will be the constants 10 and 2, respectively. A similar discussion using row 7 should be enough to convince students that the cells in columns E and I should be expressions. At this point in the discussion, several different values should be entered into the cells in column C to gain insight into the problem and to obtain some good approximate solutions to the original problem.

After this experience the class should devise an algebraic model of the problem (see fig. 22.2, row 12). The algebraic model should then be used to determine an exact answer. This answer can be used to evaluate the quality of the approximate solution attained by means of the spreadsheet model.

The primary benefits of using a spreadsheet program lie in its ability to do accurate calculations quickly and to provide a visual record of the results of all explorations with the model. In addition, once the model is set up, it can be easily modified to explore variations.

One critical part of using a spreadsheet model with students is involving them in the process of building the model. This in itself is a process through which insight into a problem and into the nature of its solution may be gained. Students should not just be handed a spreadsheet model (template) and told to use it as a computational shortcut.

Several additional topics lend themselves to exploration using spreadsheet models. Nearly any problem that allows tabular representation is appropriate for consideration as a spreadsheet application. Time, rate, and distance problems would qualify, as well as the iterative solution of a system of equations.

23

Using Computer Graphing Software Packages in Algebra Instruction

Joyce S. Friske

COMPUTER graphing software packages are instructional tools that have the potential to enhance, reinforce, and build algebra graphing concepts and skills. Using this type of software, the classroom teacher can design instructional environments that ask the student to conceptualize graphical representation, to manipulate functions and expressions, to explore graphs of functions, and to solve graphing problems. The flexibility and speed of the computer in displaying information graphically allow all this to happen without the energy and time usually spent by teachers and students in "plotting points and drawing the graph."

GENERAL DESCRIPTION OF THE SOFTWARE

Many graphing packages are available commercially. The following discussion describes characteristics found in many packages. The reader should test the software prior to purchase to see that it does what is desired.

Computer graphing software can be classified into two categories: (*a*) function graphers, and (*b*) graphing games. Many packages provide both options for instructional use. Graphing software is designed for easy access and use by teachers and students. Most are designed to be useful in any algebra classroom, since they usually are not text specific.

Function graphers allow the student or the teacher to select a graph type, such as a parabola, a circle, or a line, to be graphed on the computer screen when an appropriate equation is entered from the keyboard. Usually the coordinate system will be displayed also. The strength of the technology is that the key parameters in the defining equation are entered by the user and the graph is drawn rapidly with little or no user calculation. For example, in

181

graphing linear functions, the computer will request the values of m and b in $y = mx + b$ or the values of A, B, and C in $Ax + By + C = 0$. This allows the user to *vary* these parameters to help learn what effect they have on the graph. Most function graphers also permit multiple graphs to be displayed on the same coordinate system. This permits visual comparisons of graphs with different defining equations.

Graphing games come in a variety of formats relating to various parts of the algebra curriculum. One example, *Green Globs* (Dugdale 1986), asks the user to explode a set of randomly placed "globs" displayed on a coordinate system by "hitting" them with graphs of equations entered at the keyboard. Another game requires the user to identify the type of graph displayed and to enter its equation at the keyboard, whereas another involves identifying a function by determining the coordinates of points on the graph where known linear functions intersect.

IDEAS FOR USE

Function Graphers

As a demonstration tool, function graphers can be used to illustrate or guide concept development during instruction. As the graphs of specific functions are studied, the class can explore "What if . . ." questions with the outcomes quickly observed—for example, questions like "What happens if we change the 2 in $2x + 3y + 5 = 0$ to a -2?" and "How are the graphs related?"

Teachers can use function graphers to teach specific concepts in algebra. For example, to study the concept of slope, the function grapher can be used to illustrate both positive and negative slopes. A variety of different slopes can be illustrated by having the computer draw many different lines on the same screen. Students can be asked to predict what the next line will look like or to give an equation of a line with a particular slope. The computer's speed in graphical display allows quick and accurate checks. The machine can be used to show the relation of two lines with the same slope or with negative reciprocal slopes or with opposite slopes, all in almost no time at all.

Function graphers can also be used in review and evaluation. The teacher can enter a graph and ask the students to identify its essential characteristics. For example, does the parabola have a maximum or a minimum, is the leading coefficient positive or negative, is the constant term positive or negative, and so on. For review, these questions would guide class discussion; for evaluation, each student would answer independently.

Function graphers can be used for individual investigation and experimentation. Some packages include in the documentation worksheets or investigation suggestions that can be used by individual students to review or extend their learning independently. If one is fortunate enough to have a number of computers available, individual activities can become class activities with the students working alone, in pairs, or in small groups. Teachers can supplement these commercial materials with investigations of their own design. Simply pose a question for investigation and let the kids use the function grapher to help them resolve the question.

Graphing Games

Graphing games provide a motivating environment for problem solving. Using the game as a class activity, the students can ask questions, gather data, and make predictions that can be checked by the computer. For example, while trying to find a hidden graph of a linear equation, students can develop point-locating strategies, use the data gathered to determine the equation, and generalize successful strategies for use with other linear and nonlinear graphs. If done individually, students should share their successful strategies with the remainder of the class. In another version of the "identifying the graph" activity, the graph is displayed along with a table of values. The students need to identify the class of function graphed and to supply the correct parameters to generate the given graph. This activity exercises skills of function identification not easily done without computer technology.

A major strength of graphing games is that they require using a level of algebraic knowledge *and* skill that grows with the level of the student. A good example is *Green Globs*. For example, introductory algebra students could use the graphs of the functions they know in their attempts to hit the globs—usually linear and quadratic. They would need to estimate the coordinates of the globs and to determine an appropriate equation. Advanced algebra students could experiment with graphs of polynomial, rational, logarithmic, exponential, and trigonometric functions. As students try to get the best-fitting curve, a good deal of analysis, estimation, manipulation, and synthesis is used—all goals of algebra. As new algebra functions are studied, students can expand and extend the strategies and techniques used to hit the globs. Of course, as with any activity, advanced planning for the use of graphing games is necessary. Some teachers use them as often as one day in ten.

SOFTWARE SELECTION AND CLASSROOM MANAGEMENT

Graphing software packages are available primarily for the Apple II and IBM families of computers. The selection of a graphing program should depend on many factors including teaching style, availability of hardware, and the goals of the algebra curriculum. From a list of software recommended for previewing in the annual publication of the Educational Software Evaluation Consortium (1987), the teacher should obtain preview copies, examine the packages thoroughly, and decide on the most useful package for the local situation.

Using graphing software to teach algebra should be planned to complement the existing curriculum and the teacher's teaching style. The teacher should become thoroughly familiar with the software to develop confidence with its operation. The amount of time needed for effective use of these programs varies greatly. The use of some packages will need to be explained and demonstrated to the students initially. In addition, the teacher will sometimes need to design student guidesheets to be used in individual activities to ensure attainment of the objectives.

Classrooms with a single computer can employ group problem-solving activities over a number of months when using the graphing games. A bulletin-board display of group attainments could be kept. Multicomputer classrooms or laboratories can support small-group activities or individual assignments that can be accomplished in a daily lesson or over an extended time.

Initially the time commitment necessary to use any computer graphing package in the classroom may be greater than the time typically devoted to teaching the same material. However, the active participation of the students in the process seems to negate this drawback. Graphing software packages have the potential to enhance students' understanding of important aspects of algebra such as variable, function, and multiple representation of concepts. In addition, motivated algebraic manipulation will be done in working on the graphing games. These positives seem to support the use of computer graphing packages in the teaching of the school algebra curriculum. Try them for yourself!

REFERENCES

Educational Software Evaluation Consortium. *1987 Educational Software Preview Guide.* Sacramento, Calif.: California State Department of Education, 1987.

Dugdale, Sharon, and D. Kibbey. *Green Globs and Graphing Equations.* Pleasantville, N.Y.: Sunburst Communications, 1986.

CAN YOUR ALGEBRA CLASS SOLVE THIS?

Problem 19. Find the value(s) of $x + y$ if $x^2 + y^2 = 36$ and $xy = -10$.

Solution on page 248

24

Logarithms, Calculators, and Teaching Intermediate Algebra

Betty J. Krist

THROUGHOUT history mathematics has been investigated by observation and quasi-experiments. The recent proof by deBranges of the sixty-eight-year-old Bieberbach conjecture (concerning the class of one-to-one analytic functions, f, defined on the unit disk in the complex plane, for which $f(0) = 0$ and $f'(0) = 1$) illustrates this method. "Though his final proof is independent of it, deBranges credits a large computer at Purdue University with an important role in his discovery. . . . [He] worked with Walter Gautschi of the computer science department at Purdue to check special cases. The promising results encouraged deBranges to continue his ultimately successful line of attack on the general conjecture" (Zorn 1984).

Now the calculator can provide a laboratory for students to investigate mathematics. They can perform experiments, develop their own ideas, and use the strategies of a natural scientist. Experimentation can be a routine tool in both teaching and learning, but skills must be honed because many teachers and students have little experience in this style of mathematical thinking.

How can a teacher present standard curriculum content to students by means of exploring data and ideas? Here is a collection of lessons that have been used with different heterogeneous classes of eleventh-grade students, each time with the same dramatic results. These are activity-oriented lessons for intermediate algebra students studying exponents. Each student had been issued a scientific calculator and specific instruction in how to use it. Classroom notes follow.

LESSON 1

The class began with a simple direction: "Your calculator has a key labeled LOG. What does this key do?" One student was chosen to be secretary to write notes on the board. The teacher merely took a seat in the back of the room. (Be careful! This role is much harder than you might think.) The students agreed to fix their calculators to display four decimal places and began by making the table in figure 24.1. Different students tried different values of n in log n and reported to the secretary.

n	log n	n	log n
0	error	−1	error
1	0.0000	−2	error
2	.3010	100	2.0000
8	.9031	1000	3.0000
3	.4771	10000	4.0000
4	.6021	30	1.4771
10	1.0000	300	2.4771
5	.6990	3000	3.4771
9	.9542	50	1.6990
6	.7782	500	2.6990
7	.8451	5000	3.6990

Fig. 24.1

At this point two students made comments: "This is not the real thing— it's rounded—3 should be .477121255, but even that's probably rounded." "This is getting us nowhere; we should look for some patterns." With this, the teacher suggested that students make a list of conjectures and look for relationships within the table. The students began by making the following list of conjectures:

Student 1: $10^{\log n} = n$ $10^{.6021} = 4.004$

$10^{.3010} = 1.9999$ $10^1 = 1$ $10^2 = 100$

(This rather startling *initial* conjecture, $10^{\log n} = n$, was the first to be expressed in *each* of the classrooms where this lesson was presented. This may be attributed in part to the fact that LOG and 10^x are both written on the same calculator key and that this lesson was part of a unit on exponents.)

Student 2: 5, 50, and 500 have the same decimal (in the log) and the same number in front of the zeros.

Student 3: That also works for 1, 10, 1000; 3, 30, 300.

Student 4: But .5 doesn't work like the 5's.

Student 5: But .2 extends the 5's series; .2 yields .6990, so there is a relation between 2's and 5's.

Student 6: Negative logs give (the number of) zeros in front:

$$.5 \rightarrow -.3010$$
$$.05 \rightarrow -1.3010$$
$$.005 \rightarrow -2.3010$$

Student 7:   log 5 + log .2 = 0

log 2 + log .5 = 0

log n + log $(1/n)$ = 0

Student 8:   log 5 − log .5 = 1

log 2 − log .2 = 1

log n − log $.n$ = 1

|log n| + |log $.n$| = 1

Student 9: The number before (the log) is like the exponent in scientific notation except for negatives—it's one less than the number of digits.

The class discussion stopped and the teacher decided to intervene by asking students to consider the table in figure 24.2 and asking them for any relations among the numbers in this table.

n	log n
1	0.0000
2	.3010
3	.4771
4	.6021
5	.6990
6	.7782
7	.8451
8	.9031
9	.9542
10	1.0000

Fig. 24.2

Students again ignited, producing these comments:

2 and 3 give 6   .3010 + .4771 = .7782 almost
2 and 4 give 8   .3010 + .6021 = .9030 almost
3 and 3 give 9   .4771 + .4771 = .9542 exactly

The teacher asked students to complete the following table as well as they could *without* using their calculators:

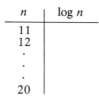

n	log n
11	
12	
.	
.	
20	

Students made the entries shown in figure 24.3. The teacher asked the students to check the logs by using their calculators. Discrepancies were found. With this the class period was drawing to a close, so the teacher assigned homework: "Look at our tables and conjectures and try to find errors or decide why our statements are true."

n	$\log n$	
11	1.0396	halfway between 10 and 12
12	1.0792	3 and 4 or 2 and 6
13	1.1127	halfway between 12 and 14
14	1.1461	2 and 7
15	1.1761	3 and 5
16	1.2041	2 and 8 or 4 and 4
17	1.2297	halfway between 16 and 18
18	1.2552	2 and 9 or
	1.2553	3 and 6
19	1.2782	halfway between 18 and 20
20	1.3010	10 and 2 or
	1.3011	4 and 5

Fig. 24.3

LESSON 2

The next day's class opened with a smug pronouncement from one student: "The key is the first statement,

$$10^{\log n} = n.$$

The rest is just how exponents work. The errors are because of rounding." When asked about the logs of 11, 13, 17, and 19, he said, "Exponents don't have even jumps: $2^2 = 4, 2^3 = 8, 2^4 = 16$. The differences are not the same."

Using the statement $\log x = a$ **iff** $10^{\log x} = 10^a = x$, students spent the remainder of this second class proving the usual theorems about logarithms:

$$\log (xy) = \log x + \log y$$
$$\log (x/y) = \log x - \log y$$
$$\log x^p = p \log x$$

The teacher then wrote the following question on the board:

Find n to the nearest hundredth so that

(a) $4^n = 5$

(b) $4^n = 19$

(c) $4^n = 95$

She circulated around the class and made some helpful comments but didn't say too much. When most of the students had just about finished, she said, "Oh! I forgot to write all of the question on the board. Here is the rest of it":

(d) Complete the table

n	4^n
	5
	19
	95

(e) Go back and do (a), (b), and (c) *another* way.

Students groaned, since they had generally solved the first three parts of this problem by a guess-and-check technique using their calculators, and they had probably considered the equations independent of each other and of the theorems that had just been proved. But when they completed the chart

n	4^n
1.161	5
2.124	19
3.285	95

and turned to part (e), they started to put these ideas together.

This lesson again requires the teacher to be unobtrusive and patient. Students need an opportunity to talk to each other and time to do some independent thinking. Given such a setting, most students will verify for themselves that if $x^n = y$, then $n = (\log y)/(\log x)$. It is also worth noting here that calculator-equipped students can easily make the transition from common logarithms to natural logarithms and to logarithms having any positive real base. This point in the lesson is a good time to show and prove the following:

If $x^n = y$, then $n = (\log y)/(\log x) = (\ln y)/(\ln x) = (\log_b y)/(\log_b x)$, b a positive real. Teachers are encouraged to deal with natural logarithms and e in these lessons, but their own good judgment must prevail.

These homework questions were assigned:

1. Find n to the nearest thousandth so that $2.1^n = 7$.
2. Solve for x to four significant digits and check:

    a. $5 \cdot 3^{2x+1} = 11^{1-x}$

    b. $3^{x+2} = 5^{x-2}$

3. How does the solution of $2^{49} = 10^x$ verify that 2^{49} has fifteen digits?

LESSON 3

This lesson focused on the graphs of exponential and logarithmic functions, which can easily be sketched by calculator-equipped students. They observed some important properties and relationships of these functions, and the sketches they produced not only seemed to be more easily remembered but also made a worthwhile bulletin board.

The classroom activity was to have ten different groups of students each sketch one of the following:

$$f(x) = 2^x \qquad k(x) = \log_2 x$$
$$g(x) = 3^x \qquad l(x) = \log_3 x$$
$$h(x) = 2.3^x \qquad m(x) = \log_{2.3} x$$
$$i(x) = 3.1^x \qquad n(x) = \log_{3.1} x$$
$$j(x) = e^x \qquad p(x) = \ln x$$

The students were not given a specific domain, but each group worked with the same axes scale. Their work was checked by the teacher, and then they traced their graphs with various-colored markers on acetate to be used on an overhead projector. Properties of exponential and logarithmic functions were then easily demonstrated:

For $f(x) = b^x$, $b > 1$, b constant—
$f(0) = 1$;
as $x > 1$ increases, $f(x)$ increases without bound;
as $x < 0$ decreases, $f(x)$ decreases and is bounded by 0;
the domain of $f(x)$ is all real numbers;
the range of $f(x)$ is all positive real numbers.

For $g(x) = \log_b x$, $b > 1$, b constant—
$g(1) = 0$;
as $x > 1$ increases, $g(x)$ increases without bound;
as $0 < x < 1$ decreases, $g(x) < 0$ and decreases without bound;
the domain of $g(x)$ is all positive real numbers;
the range of $g(x)$ is all real numbers.
$f(x)$ and $g(x)$ are inverse functions.

The homework to accompany this lesson was as follows:

1. Sketch the function $f(x) = \left(\dfrac{1}{2}\right)^x$ for values of x between -3 and 3 inclusive. Label the graph with its equation.

2. On the same set of axes used in part 1, graph the function $g(x) = \left(\dfrac{1}{2}\right)^{-x}$ for values of x between -3 and 3 inclusive. Label the graph with its equation.

3. Write a function that is the inverse of $f(x) = \left(\dfrac{1}{2}\right)^x$.

4. Does the function $g(x) = \left(\dfrac{1}{2}\right)^{-x}$ have an inverse? If so, what is it? If not, why not?

CONCLUSIONS

Several observations can be made about these lessons. First, genuine discovery activities are exciting. The students were enthusiastic about their work. They acted as detectives. They enjoyed the clues given by their calculators. No one said, "Just tell us how it works." It would seem they sensed that the teacher would have spoiled their fun if all questions had been answered immediately as they arose. The students obtained the important properties about logs and why they were true by themselves. They even discovered a few rather obscure properties:

$$|\log n| + |\log (n/10)| = 1, 1 \le n \le 10$$

Second, the calculator provided a framework for the activity and helped the students put *their own* ideas together. These activities would be hard to imagine in a classroom with tables rather than calculators. The calculators provided a setting and were an aid to student thinking, but the students had to go beyond the numbers that were exhibited in the calculator's display.

Lastly, the calculators allowed for an instructional style that is far less pedantic than that usually associated with the teaching of logarithms. The teacher and students had an opportunity to be engaged in serious dialogue about mathematics, to be concerned about the nature of inquiry and proof, and to actually do mathematics.

REFERENCE

Zorn, Paul. "Bieberbach Conjecture Proved." *Focus* [newsletter of the Mathematical Association of America] 4 (November-December 1984):3, 7.

CAN YOUR ALGEBRA CLASS SOLVE THIS?

Problem 20. A merchant paid $30 for an article. He wishes to place a price tag on it so that he can offer a 10% discount on the price marked on the tag and still make a profit of 20% on the cost. What price should he mark on the tag?

Solution on page 248

CAN YOUR ALGEBRA CLASS SOLVE THIS?

Problem 21. When 6 gallons of gasoline are put into a car's tank, the indicator goes from 1/4 of a tank to 5/8. What is the total capacity of the gasoline tank?

Solution on page 248

25

Factoring Twins as a Teaching Tool

Lee H. Minor

EVERY student in a beginning algebra course has probably shown that each of the trinomials $x^2 + 5x \pm 6$ has linear factors with integer coefficients. In particular,

$$x^2 + 5x + 6 = (x + 2)(x + 3)$$
$$x^2 + 5x - 6 = (x + 6)(x - 1).$$

Teachers and more experienced students recognize these examples to be somewhat exceptional. If two quadratics differ only in the sign of the constant term and one of them factors over the integers, the other usually does not.

USING FACTORING TWINS

I refer to $ax^2 + bx \pm c$ as *factoring twins* if both quadratics have linear factors with integer coefficients. Some examples with relatively small coefficients are shown in the box below.

$x^2 + 5x \pm 6$	$x^2 + 17x \pm 60$	$x^2 + 25x \pm 150$
$x^2 + 10x \pm 24$	$x^2 + 25x \pm 84$	$x^2 + 41x \pm 180$
$x^2 + 13x \pm 30$	$x^2 + 20x \pm 96$	$x^2 + 29x \pm 210$
$x^2 + 15x \pm 54$	$x^2 + 26x \pm 120$	$x^2 + 37x \pm 210$

Additional examples can be obtained from this list by changing the signs of the middle coefficients and also by using the fact that both of $ax^2 + bx \pm c$

factor if and only if $x^2 + bx \pm ac$ also factor. Another way to generate more twins will be given later.

The goal of this article is to illustrate some ways that factoring twins (for brevity, FTs) might be helpful to teachers in the algebra classroom. One nice feature of FTs is that changing the sign of *any* coefficient in a twin results in a quadratic that also has linear factors. Thus a factoring problem involving a twin isn't ruined by an incorrect sign caused by student or teacher. As might be expected, FTs can be a useful tool when teaching techniques related to factoring itself.

Consider $x^2 + 10x - 24$ and $x^2 + 10x - 39$. Will either of these reveal more about a student's ability to factor trinomials than the other? My experience has been that those who approach factoring impulsively or treat it as guesswork are more likely to incorrectly factor a twin than a nontwin with the same middle coefficient. This is probably because the middle coefficient of a twin can always be obtained in two ways using relevant factor pairs; this is not true for nontwins. By illustration from the quadratics above, $10 = 6 + 4 = 12 - 2$, as opposed to $10 = 13 - 3$.

When the factors of a trinomial are not readily apparent, I try to discourage a trial-and-error search for them by encouraging a methodical approach such as the "*ac* method" for factoring $ax^2 + bx + c$. Here are three nonroutine problems that can be done using this method.

Problem 1

Determine a value for b that makes FTs of the following, and find the linear factors of each twin: $2x^2 + bx \pm 27$ and $x^2 + bx \pm 36$. The inefficiency of a trial-and-error approach is obvious. But by listing the factor pairs of the ac product and their sums and differences, we find that the problem becomes very straightforward. The desired value of b is simply the number (if any) that appears as both a sum and a difference.

For $2x^2 + bx \pm 27$ we have $2 \cdot 27 = 54$, so consider the table:

54	Sum	Difference
$54 \cdot 1$	55	53
$27 \cdot 2$	29	25
$18 \cdot 3$	21	$\boxed{15}$
$9 \cdot 6$	$\boxed{15}$	3

Thus $b = 15$, so the *FTs* are $2x^2 + 15x \pm 27$. Then:

$$\underline{2x^2 + 15x + 27} \qquad\qquad \underline{2x^2 + 15x - 27}$$
$$= 2x^2 + (9x + 6x) + 27 \qquad = 2x^2 + (18x - 3x) - 27$$
$$= (2x^2 + 9x) + (6x + 27) \qquad = (2x^2 + 18x) + (-3x - 27)$$
$$= x(2x + 9) + 3(2x + 9) \qquad = 2x(x + 9) - 3(x + 9)$$
$$= (2x + 9)(x + 3) \qquad\qquad = (x + 9)(2x - 3)$$

For $x^2 + bx \pm 36$ consider the table:

36	Sum	Difference
$36 \cdot 1$	37	35
$18 \cdot 2$	20	16
$12 \cdot 3$	15	9
$9 \cdot 4$	13	5
$6 \cdot 6$	12	0

Since no number appears as both a sum and a difference of two factor pairs of 36, there can be no *FT*s of the form $x^2 + bx \pm 36$.

Problem 2

Find every positive value of b for which the trinomials $2x^2 + bx \pm 27$ have linear factors. Determine the factors of each such trinomial, and identify any factoring twins that occur.

The desired values of b are just the sums and differences from the preceding table of factor pairs of 54. The factorings follow:

$2x^2 + 55x + 27 = (2x + 1)(x + 27)$ $2x^2 + 53x - 27 = (2x - 1)(x + 27)$

$2x^2 + 29x + 27 = (2x + 27)(x + 1)$ $2x^2 + 25x - 27 = (2x + 27)(x - 1)$

$2x^2 + 21x + 27 = (2x + 3)(x + 9)$ $2x^2 + 15x - 27 = (2x - 3)(x + 9)$

$2x^2 + 15x + 27 = (2x + 9)(x + 3)$ $2x^2 + 3x - 27 \;\; = (2x + 9)(x - 3)$

Solving the main problem requires no knowledge of FTs, of course, and twins will not occur on every problem of this type.

The beginning student may expect many more answers if confronted with this kind of problem before seeing an example from an instructor. Thus it not only provides a lot of factoring practice but also illustrates that factoring possibilities are not endless.

Problem 3

Factor each of the following trinomials using the factor pairs of 60:

$x^2 + 17x + 60$ $3x^2 + 17x + 20$ $5x^2 + 17x + 12$

$2x^2 + 17x + 30$ $4x^2 + 17x + 15$ $6x^2 + 17x + 10$

Because each of these has an ac product of 60, an obvious advantage for the student is the opportunity to factor all of them using the same table of sums and differences. Because each of these is also a twin, an advantage for the teacher is that forty-eight possible problems are available when all variations in signs of all the coefficients are considered.

We now turn to solving some equations and inequalities (problems 4–12) in which FTs can be used to produce interesting and instructive variations in some problems that otherwise become rather monotonous:

4. $| x^2 - 5x | = 6$ 7. $| x^2 - 5x | < 6$ 10. $\sqrt{6 - 5x} = x$

5. $| x^2 - 6 | = 5x$ 8. $| x^2 - 6 | < 5x$ 11. $\sqrt{6 - 5x} \geq x$

6. $| 5x - 6 | = x^2$ 9. $| 5x - 6 | \geq x^2$ 12. $\sqrt{6 - 5x} < x$

Problem 4: $| x^2 - 5x | = 6$

Nothing is quite so boring and predictable to us as solving absolute-value equations of the form $| ax + b | = c$. Even students who do not understand absolute values can often do them, except when we slip in a negative value for c. A variation like this, with four solutions instead of two, may stimulate some thought in students' minds about what is really involved. The given problem is equivalent to

$$x^2 - 5x = 6 \quad \text{or} \quad -(x^2 - 5x) = 6$$
$$x^2 - 5x - 6 = 0 \quad \text{or} \quad x^2 - 5x + 6 = 0$$
$$(x - 6)(x + 1) = 0 \quad \text{or} \quad (x - 2)(x - 3) = 0$$

Thus the solutions are $-1, 2, 3$, and 6. Coefficients of FTs aren't needed, but their use assures that both quadratics will factor.

Teachers who like to relate solving equations and inequalities to graphs of functions should enjoy this problem. Indeed, a consideration of graphs here provides insight into why there are four solutions. For $c > 0$, figure 25.1a demonstrates that the graphs of $y = | ax + b |$ and $y = c$ must always intersect in two points, thus ensuring that $| ax + b | = c$ has two solutions. Figure 25.1b illustrates the functions associated with problem 4. We can see that when $c > 0$, the graphs of $y = c$ and $y = | ax^2 + bx |$ will intersect in two, three, or four points, depending on the particular value of c.

Problems 5, 6: $| x^2 - 6 | = 5x$ $| 5x - 6 | = x^2$

These problems show that when using FTs, "nice" answers are still obtained if terms in the preceding problem are permuted. The solutions follow

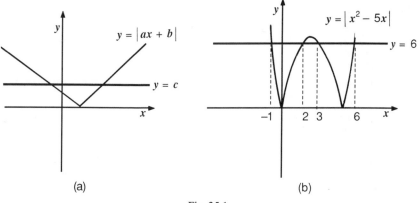

Fig. 25.1

in a manner similar to those for problem 4, with one important difference. Notice in problem 5 it is necessary that $x \geq 0$, since $5x$ is equal to an absolute value. So the solutions for problem 5 are 1 and 6 only. Problem 6 involves no such restrictions; its solutions are $-6, 1, 2$, and 3. The graphs associated with these problems appear in figure 25.2.

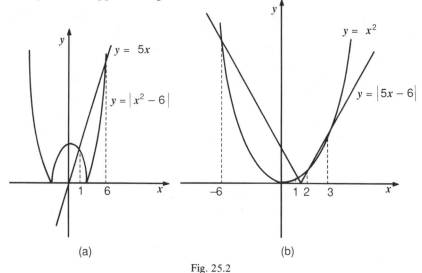

(a)   (b)

Fig. 25.2

Problem 7: $|x^2 - 5x| < 6$

Rather than work this problem directly by solving the inequalities in $-6 < x^2 + 5x < 6$, we prefer the "test value" approach, using the fact that solutions to the corresponding equation (problem 4) separate the solutions of the inequality from the nonsolutions. Thus we choose values from each of the open intervals separated by solutions to the equation and test them by substitution into the inequality. For this problem we might have the following:

Interval	Test Value	Substitution	Conclusion
$(-\infty,-1)$	-2	$14 < 6$	Not a solution
$(-1,2)$	0	$0 < 6$	Solution
$(2,3)$	2.5	$6.25 < 6$	Not a solution
$(3,6)$	4	$4 < 6$	Solution
$(6,+\infty)$	7	$12 < 6$	Not a solution

We conclude that the solution set consists of the union of the intervals $(-1,2)$ and $(3,6)$. This is also apparent from the graphs in figure 25.1b.

Problems 8, 9: $|x^2 - 6| < 5x$ $|5x - 6| \geq x^2$

These may be done directly, or by reference to the corresponding equations (problems 5 and 6, respectively), or by reference to the graphs in figure 25.3. The solution set for problem 8 is the open interval (1,6). For problem 9 the solution set is the union of closed intervals $[-6,1]$ and $[2,3]$.

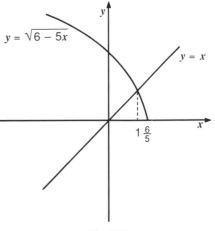

Problem 10: $\sqrt{6 - 5x} = x$

There are no surprises in this problem. We must have $6 - 5x \geq 0$ to obtain real solutions, so the equation is meaningful only if $x \leq 6/5$. Squaring both sides leads to $(x + 6)(x - 1) = 0$, but only 1 is a solution.

Fig. 25.3

Using coefficients from FTs again allows some flexibility. The quadratics that arise are factorable, and equations like $\sqrt{5x - 6} = x$ and $\sqrt{6 - 5x} = x$ might be used as parallel problems on different versions of a test. Notice, however, that to maintain integer coefficients, only FTs with perfect square coefficients of the x^2 term can be considered. Thus, $4x^2 + 17x + 15$ can be adapted to the equation form ($2x = \sqrt{17x + 15}$) but not $2x^2 + 17x + 30$. It should also be noted that the terms can't be permuted as freely as in problems 4–9.

Problems 11, 12: $\sqrt{6 - 5x} \geq x$ $\sqrt{6 - 5x} < x$

We can use the results from problem 10 or the graphs shown in figure 25.3 to work problems 11 and 12. The inequality in problem 11 is meaningful only if $x \leq 6/5$ (which is where both graphs exist in fig. 25.3). Using the test-value approach, we see that the intervals to be tested are $(-\infty,1)$ and $(1,6/5]$. The solution set is $(-\infty,1]$.

Notice that problem 12 requires that $x > 0$, since x is larger than a square root, which cannot be negative. Thus we need only consider this inequality on the interval $(0,6/5]$. The solution set $(1,6/5]$ can be obtained with test values or by referring to figure 25.4. We could also observe that problems 11 and 12 differ only in the direction of the inequalities, so their solution sets must be complementary relative to the interval $(-\infty,6/5]$ on which both are defined.

EXTENSIONS

Teachers who find some appeal in the concept of factoring twins may discover additional applications during their next algebra course. FTs also have some properties that should stimulate the curiosity of anyone with an interest in elementary number theory. Thus, they could be used as enrichment material for some better students. A few concluding remarks may encourage further study by the interested reader. For more complete details, consult Minor (forthcoming).

Quite often mathematical conjectures are formulated on the basis of observing a recurring pattern in a number of examples. FTs provide such an opportunity. Careful study of the twelve examples in the box at the beginning of this article suggests some interesting possibilities.

For example, if $x^2 + Mx \pm N$ are FTs, then it appears that—

1. $x^2 + pMx \pm p^2N$ are FTs for any integer $p \neq 0$;
2. M must be less than N;
3. N must be divisible by 6.

Each of these is indeed true! To show (1), we merely observe that if $x^2 + Mx + N = (x + I)(x + J)$ and $x^2 + Mx - N = (x + L)(x - K)$, then $x^2 + pMx + p^2N = (x + pI)(x + pJ)$ and $x^2 + pMx - p^2N = (x + pL)(x - pK)$. Thus from any pair of twins an infinite number of additional twins can be generated.

The second property above is easily verified. The third is more difficult but can be done by proving that both 2 and 3 must divide N. However, (3) follows easily from another rather surprising result. It can be shown that M must be the hypotenuse of a Pythagorean right triangle whose area is N! Since it is known that the product of the lengths of the legs of a Pythagorean right triangle is divisible by 12, the area (being half that product) is divisible by 6.

Writing a computer program to find FTs can be an interesting experience. Teachers might want to challenge some of their students to find answers to the following (with or without a computer):

If $x^2 + Mx \pm N$ are FTs,

1. What is the smallest value of N that can occur in two different FTs? In three different FTs?
2. What is the smallest value of M that can occur in two different FTs? In three different FTs?
3. For a given N, how many values of M are possible?
4. For a given M, how many values of N are possible?

REFERENCE

Minor, Lee H. "When Twins and Triplets Are Equivalent." Forthcoming.

26

An Algebra Class Unveils Models of Linear Equations in Three Variables

Deborah Davies

THIS article describes a unit for an honors second-year algebra class. The usual material involving systems of linear equations and inequalities in two variables is expanded to include a substantive study of systemssof linear equations in three variables. The culminating task of the unit is building a three-dimensional model. The rationale for choosing this hands-on activity is that students need to develop spatial-visualization skills, as well as algebraic skills, to be successful in calculus and other mathematics and science courses. In the standard high school curriculum for our very brightest students, too often the focus is on the abstract as students surge ahead to cover the material so they can take Advanced Placement Calculus in high school.

THE GROUNDWORK

Students begin by plotting points in three-space. The front left-hand corner of the classroom is selected to be $(0, 0, 0)$ so that the axes can then be easily assigned to produce a right-handed coordinate system. We use one meter to be one unit on each axis, and students quickly take to plotting points with their coordinates, in all eight octants, all over the school and its grounds. Nothing is immune from their assault: they delight in taping an index card marked with the corresponding coordinates to the telephone on the principal's desk, the ceiling of our classroom, the wastebasket in a nearby classroom, the floor in a room two flights down, and so on.

Next, time is spent discovering that the general linear equation in three variables, $Ax + By + Cz = D$, has a graph that is a plane, not a line. We begin by graphing $x = 0$ and the other coordinate planes.

Then students tackle planes parallel to the coordinate planes. By now we have written the equation for each of the six planes in the classroom on an

index card and taped the cards to the planes. The walls of nearby classrooms and halls are targets for more index cards with equations of those planes. Students have an easy time graphing equations with nonzero coefficients, such as $2x - 5y + 3z = 60$, by plotting the three intercepts and drawing the "triangle" formed by the three traces of the plane in the coordinate planes. Planes corresponding to equations that have zero as one of the coefficients of the three variables are harder to visualize immediately. At this point the teacher is wise to discuss the similarities between graphing in two-space and three-space. For example, an equation in two-space with a "missing" variable x, such as $y = 6$, results in a line parallel to the x-axis. In the same way, the graph of an equation in three-space with a "missing" variable x, such as $2y + 3z = 6$, results in a plane parallel to the x-axis. These analogies are extremely helpful by enhancing students' understanding of graphing in both two and three dimensions.

Teachers need to handle with care the special case when $D = 0$. Students become dependent on using the intercepts as starting points for the graphs, and when all the intercepts are zero, the problem becomes particularly difficult. Again, the analogy to two-space is very helpful. Students have an easier time with $D = 0$ if we arbitrarily move the origin from the corner of the room to a more accessible location, equip them with metersticks to represent the axes, find a few points on the plane, and then have them explain where the plane should be.

THE PAPERWORK

The graphing of a single plane is extended to graphing several linear equations in three variables. The graph of these equations is the union of planes, a three-dimensional surface, which the students visualize as a hollow "solid." The three coordinate planes are graphed as part of the equation set in order to restrict the graph to the first octant. We spend a few days graphing three-dimensional surfaces. Drawing these surfaces is a most difficult task for students. It is less threatening and more productive if they work on the assignment in small groups. Often those students with a background in artistic drawing fare better than the others. The ability to visualize things in three dimensions is a skill our best mathematics students do not necessarily possess! Solving for the vertices of the surface automatically leads to solving systems with three unknowns, and the standard methods—including linear combination, substitution, determinants, and matrices—are presented. For example, the set of equations below has the graph shown in figure 26.1.

$$x = 0$$
$$y = 0$$
$$z = 0$$
$$x = 7$$
$$2y + 3z = 6$$

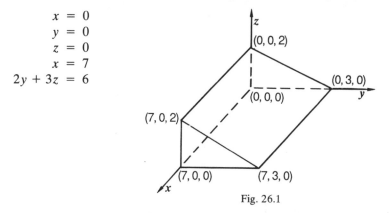

Fig. 26.1

Drawing the surfaces forces students to use and understand the postulates from geometry involving points, lines, and planes, They make considerable use of the following: "Through any two points there is exactly one line"; "If two points are in a plane, then the line that contains the points is in that plane"; "If two planes intersect, then their intersection is a line"; and "Three noncollinear points determine a plane."

THE EXAM

After the students have become proficient with the needed skills and concepts, they are given a lengthy take-home exam to be done individually. This exam covers linear programming, plus the usual topics associated with a study of systems of linear equations and inequalities. The heart of the exam is a problem in which students are to draw a surface carefully, draw each face of the surface accurately, and construct a three-dimensional model. An example of such a problem follows:

1. Draw a sketch of the surface enclosed in the $(+,+,+)$ octant by the coordinate planes, $2y = 15$, $z = 6$, $10x + 12y + 15z = 180$ and $30x - 18y = 135$. Label the axes and the vertices with their coordinates.

2. Make an accurate two-dimensional drawing of each face of the surface. On each of your drawings, do the following:

   a) Mark the length of each side. Leave your answers accurate to two decimal places. (*Hint:* You will need to make liberal use of the Pythagorean theorem.)

b) Mark each angle with its degree measure. Your answer must be accurate within two degrees. (*Hint:* You will need to use a protractor in several places. If you have already studied trigonometry, you should be able to find much more accurate measures of the angles.)

3. Construct a three-dimensional model of the surface in part 1. Neatness, creativity, and accuracy will be factors in the grading.

The sketch of the surface should be as shown in figure 26.2.

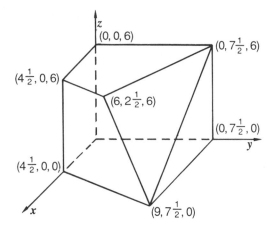

Fig. 26.2

THE MODEL

The results of this method have been phenomenal! Almost all students like this project and spend a lot of time on their models. Students have made spectacular models out of balsa wood and rice paper, gingerbread, cheddar cheese, Plexiglas, poster board or cardboard with interesting decorations,  playing cards, and many other unusual materials. Some models have come complete with trapdoors that show the inside with the equations of each plane marked. Another model had a trapdoor and a miniature of the model inside. The variety in the types of models created has been a major factor in students' overwhelming support of this project.

However, in addition to the success of these creative endeavors, some remarkable spatial problems have come to light. About one student in twenty-five will bring in a model that is very flat near the origin. This problem seems to stem from the way we draw three axes in two-space. These students visualize a two-dimensional plane with two 135-degree angles and one right angle instead of three right angles! Their reverting to two-space is incredible in light of the amount of time we spend discuss-

ing three-dimensional surfaces in a classroom, which is itself a three-dimensional model. A more common problem, one that most students encounter, is constructing the sides of the surface without first ensuring that all sides face toward either the inside or the outside of the model. Of course, I never warn them ahead of time about these potential pitfalls. It's better for them to experience them!

THE "UNVEILING"

This project has grown and prospered over the years. Students from past years are always eager to see what this year's crop of students will do for their models. We hold a presentation called the "Unveiling of the Models," to which faculty members and parents are invited. Past students request permission to attend. The parents have become very interested in these bizarre projects after seeing their teenagers suffer through the construction of a model at home. They are always fascinated to see what the other students have made. At the "unveiling" the students are asked to list the "non-mathematical" things they learned from this project. The lengthy and funny list makes everyone realize that he or she was not the only one to have at least ten major setbacks with the construction of the model.

Building models is challenging, fun, creative, very demanding, and both frustrating and rewarding for the teacher. Each year I weigh the time spent on this project against my need to march inexorably through the material. Each year I grow more firm in my conviction that more growth in my students occurs because of this project than I could wrest from them by presenting another chapter.

27

Common Mistakes in Algebra

June Marquis

A FTER a few years of teaching mathematics courses in high school, teachers know which concepts and manipulations will cause difficulty for students. From year to year, class to class, students often make the same algebraic mistakes over and over. In upper-level mathematics courses, students' indication of mastery of the new concepts may be obscured by common algebraic errors.

Following is a short test covering common errors. I found it very helpful to give the test near the end of the algebra courses, just prior to the final exams. Becoming aware of these commonly made errors, the students were less likely to make them on the final exam.

In the upper-level mathematics courses, such as trigonometry, analytic geometry, math analysis, and even calculus, I used the test on the first day of class. Even at these levels, very few students are able to answer all the questions correctly. Discussing the ones that cause the most difficulty is very beneficial. These mistakes can later be referred to meaningfully as "one of those common mistakes in algebra."

Students might be required to keep the corrected test in their notebooks for referral. Add your own favorites to supplement this list and use it in your class.

Common Mistakes in Algebra

Directions: All the statements are false. Correct each statement to make it true.

1. $|-3| = -3$

2. $3^2 \cdot 3^3 = 9^5$

3. $a^2 \cdot b^5 = (ab)^7$

4. $x + y - 3(z + w) = x + y - 3z + w$

204

5. $\dfrac{r}{4} - \dfrac{(6-s)}{2} = \dfrac{r-12-2s}{4}$

6. $3a + 4b = 7ab$

7. $3x^{-1} = \dfrac{1}{3x}$

8. $\sqrt{x^2 + y^2} = x + y$

9. $\dfrac{x+y}{x+z} = \dfrac{y}{z}$

10. $\dfrac{1}{x-y} = \dfrac{-1}{x+y}$

11. $\dfrac{x}{y} + \dfrac{r}{s} = \dfrac{x+r}{y+s}$

12. $x\left(\dfrac{a}{b}\right) = \dfrac{ax}{bx}$

13. $\dfrac{xa + xb}{x + xd} = \dfrac{a+b}{d}$

14. $\sqrt{-x}\,\sqrt{-y} = \sqrt{xy}$

15. If $2(2 - z) < 12$ then $z < -4$.

16. $\dfrac{1}{1 - \dfrac{x}{y}} = \dfrac{y}{1-x}$

17. $a^2 \cdot a^5 = a^{10}$

18. $(3a)^4 = 3a^4$

19. $\dfrac{a}{b} - \dfrac{b}{a} = \dfrac{a-b}{ab}$

20. $(x + 4)^2 = x^2 + 16$

21. $\dfrac{r}{4} - \dfrac{6-s}{4} = \dfrac{r-6-s}{4}$

22. $(a^2)^5 = a^7$

Solutions

1. 3

2. 3^5

3. $a^2 b^5$

4. $x + y - 3z - 3w$

5. $\dfrac{r - 12 + 2s}{4}$

6. $3a + 4b$

7. $\dfrac{3}{x}$

8. $\sqrt{x^2 + y^2}$

9. $\dfrac{x+y}{x+z}$

10. $\dfrac{-1}{y-x}$

11. $\dfrac{sx + yr}{ys}$

12. $\dfrac{ax}{b}$

13. $\dfrac{a+b}{1+d}$

14. $i\sqrt{x}\,i\sqrt{y} = -1\sqrt{xy}$

15. $z > -4$

16. $\dfrac{y}{y-x}$

17. a^7

18. $81a^4$

19. $\dfrac{a^2 - b^2}{ab}$

20. $x^2 + 8x + 16$

21. $\dfrac{r - 6 + s}{4}$

22. a^{10}

28

Using Polynomials to Amaze

Catherine Herr Mulligan

I N teaching algebra, I try to present the subject as relevant and useful, but I do not believe it is necessary to limit considerations of "relevance" to the real world. Most of my students will study mathematics beyond algebra 1, and I try to teach them that algebra is (1) a tool used in higher mathematics, (2) a common language, and (3) a means of communication. Real-world applications are important, but it is also good for students to see algebra used for the sake of mathematics.

Polynomial arithmetic is a good area for implementing this philosophy. Manipulating polynomial expressions is an essential skill, yet, just as with any skill requiring practice, work with polynomials can be dull and repetitious. In anticipation of students' questions, such as "Why is this important (useful) (necessary) (going to be on the test)?" I try to emphasize the value of being able to express and extend ideas.

A collection of a few "amazing facts" allows students to "discover" and then prove these interesting facts using polynomial arithmetic. This generates respect for polynomial arithmetic and demonstrates at an early level the use of algebra in proofs.

FIBONACCI PHENOMENA

Terms in a Fibonacci sequence are nicely represented by polynomials, and such a sequence is rich in patterns for students to discover and prove. First-year algebra students can deal with some Fibonacci facts early in the course when they are learning to combine like terms.

At the beginning of the course, my students are introduced to the Fibonacci sequence $1, 1, 2, 3, 5, 8, 13, 21, \ldots$, where each term is the sum of the two preceding terms. I tell them that much of mathematics consists of making observations, looking for patterns, and generalizing. They try to

206

determine the pattern in the Fibonacci sequence after they have discovered the rules for some simple arithmetic and geometric sequences.

When I write a "Fibonacci-like" sequence generated by starting with numbers other than 1's on the board, they recognize the scheme. I write, for example,

7, 11, 18, 29, 47, 76, 123, 199, 322, 521, 843, 1364, 2207.

Then I say that I will mentally sum these terms while they figure the next two terms. Students are eager to determine the new terms. After the fourteenth and fifteenth terms, 3571 and 5778, are reported, I tell them immediately that the required sum is 5767. This tidbit makes use of the following fact, adapted from a generalization found in Jacobs (1970):

Amazing Fact #1

Given fifteen consecutive terms of a Fibonacci-like sequence, the sum of the first thirteen of them is equal to the fifteenth term minus the second term.

To prove this fact, we represent fifteen consecutive terms of a Fibonacci-like sequence as the monomials and binomials: a, b, $a + b$, $a + 2b$, $2a + 3b$, $3a + 5b$, $5a + 8b$, $8a + 13b$, $13a + 21b$, $21a + 34b$, $34a + 55b$, $55a + 89b$, $89a + 144b$, $144a + 233b$, $233a + 377b$. We add the first thirteen terms and compare the sum, $233a + 376b$, to the difference of the fifteenth term and the second term, $(233a + 377b) - b$.

Students have the opportunity to see the distributive law at work when they discover, and then prove, a trick found in Gardner (1979):

Amazing Fact #2

The sum of any ten consecutive terms in a Fibonacci-like sequence is 11 times the seventh term of the ten-term sequence.

We might test this using the sequence 5, 10, 15, 25, 40, 65, 105, 170, 275, 445, 720, 1165, 1885, . . . , where we sum the ten terms beginning with 25. Multiplying 11 times 445 (the seventh term of the ten terms beginning with 25), we get 4895. Adding the terms 25, 40, . . . , 1885 gives the same result.

We prove this by noting that ten such terms could be represented by polynomials a, b, $a + b$, $a + 2b$, $2a + 3b$, $3a + 5b$, $5a + 8b$, $8a + 13b$, $13a + 21b$, $21a + 34b$. Adding these gives $55a + 88b$, which factors as $11(5a + 8b)$, and is easily identified as 11 times the seventh term.

Although this is simple, it is impressive to elementary algebra students. It promotes the idea that algebra is a tool for mathematics.

CALCULATING CURIOSITIES

"Tricks" for rapid mental computation provide many examples of phenomena that can be explained using simple polynomial representation. In this age of calculators, these phenomena are introduced, not because they are shortcuts, but because they work; students are challenged to prove *why* they work.

When students are learning the multiplication of polynomials, they are impressed with a shortcut from Fawcett and Cummins (1970):

Amazing Fact #3

If two 2-digit numbers have the same tens digit, and units digits whose sum is 10, their product can be computed instantly.

When the students test me with 77×73, for instance, I instantly reply, "5621." After another example or two, I reveal my "trick": Multiply the tens digit (7) by the next higher integer (8), getting 56, which will be the thousands and hundreds digits of the answer. Append to the right of 56 the product of the units digit: 7×3, or 21. This gives 5621.

We can increase their confidence in this procedure by trying several other examples, but many examples do not constitute a *proof.* However, if we use binomials to represent any such numbers to be multiplied, we can give a proof that does not depend on the choice of examples.

Let the common tens digit of the two numbers be represented by a. Let b represent the units digit of the first number; the units digit of the second will be $10 - b$. Then we have as the first number, $10a + b$, and as the second number, $10a + (10 - b)$. Their product is

$$(10a + b)(10a + (10 - b)) = (10a + b)(10a + 10 - b)$$
$$= 100a^2 + 100a - 10ab + 10ab + 10b - b^2$$
$$= 100a(a + 1) + b(10 - b).$$

The proof of another impressive fact involves the multiplication of polynomials that can lead to a discussion of some nonroutine factoring:

Amazing Fact #4

If you multiply any four consecutive integers and add 1, the result is always a perfect square.

A few illustrations will lead students to suspect that this assertion is always true. Our observations, recorded on the chalkboard, might appear thus:

$$1 \times 2 \times 3 \times 4 + 1 = 25 = 5^2$$
$$2 \times 3 \times 4 \times 5 + 1 = 121 = 11^2$$

If I ask for larger numbers to test the hypothesis, a student using a calculator might volunteer.

$$97 \times 98 \times 99 \times 100 + 1 = 94109401 = 9701^2.$$

At this point, the students have done textbook examples that call for algebraic representation of consecutive integers, so they are comfortable using $n, (n + 1), (n + 2), (n + 3)$ as four unspecified consecutive integers. The product, plus 1, is expanded using polynomial arithmetic:

(1)
$$n(n + 1)(n + 2)(n + 3)$$
$$= n^4 + 6n^3 + 11n^2 + 6n + 1$$

(The details of this expansion usually lead to a discussion about associativity in multiplication.) Now we proceed in one of two ways. Sometimes a student will notice that the perfect square in our numerical examples is the square of 1 plus the product of the first and last integers in the sequence. (It is also the square of 1 less than the product of the second and third integers in the sequence.)

We might observe, for example, that

$$4 \times 5 \times 6 \times 7 + 1 = 841 = 29^2 = (1 + 4 \times 7)^2.$$

Expressing this in polynomials, we write

$$(1 + n(n + 3))^2 = n^4 + 6n^3 + 11n^2 + 6n + 1.$$

Comparing this to (1) above shows that the product of four consecutive integers plus 1 is the same as the square of 1 added to the product of the first and fourth integers in the sequence. This not only confirms the fact that we have a perfect square but also tells us *what* perfect square.

Another way to proceed is to work directly from the expression (1) and speculate that it would be nice to factor it and see that it is a square. Such a square would have this form (for a suitable a):

$$(n^2 + an + 1)^2 = n^4 + 2an^3 + (2 + a^2)n^2 + 2an + 1$$

Equating coefficients with those in (1) gives

$$2a = 6 \text{ and } 2 + a^2 + 11, \text{ or } a = 3.$$

Now we know that (1) factors as the perfect square

$$(n^2 + 3n + 1)^2.$$

TRIANGULAR TRIVIA

Just recently, some older student members of our mathematics club, accustomed to hearing about amazing facts, were not surprised when I shared with them something I had come across in a journal (Demir 1986):

Amazing Fact #5

If you take the product of any four consecutive positive integers and divide by 8, the result is a triangular number.

I introduced "triangular numbers" and explained the significance of their name by listing entries in the correspondence $f(n) = n(n + 1)/2 = n(n + 1)(1/2)$: $f(1) = 1$; $f(2) = 3$; $f(3) = 6$; $f(4) = 10$; and so on.

Then we tested the hypothesis using $(3 \times 4 \times 5 \times 6)/8 = 45$ and found that 45 is $f(9)$, the ninth triangular number. For our proof, we represented the product of four positive integers divided by 8 as

$$m(m + 1)(m + 2)(m + 3).$$

Wishing to exhibit the factor $1/2$, we rewrote the expression as

$$\frac{m(m + 3)}{2} \cdot \frac{(m + 1)(m + 2)}{2} \cdot \frac{1}{2} = \frac{m^2 + 3m}{2} \cdot \frac{m^2 + 3m + 2}{2} \cdot \frac{1}{2}$$

$$= \frac{m^2 + 3m}{2} \left(\frac{m^2 + 3m}{2} + 1 \right) \frac{1}{2} .$$

Then, with comments about the manipulation of fractions, we saw that we had a triangular number, that is, a number of the form $n(n + 1)(1/2)$ where $n = (m^2 + 3m)/2$.

OTHER AMAZING FACTS

Amazing Fact #6

If a number with an even number of digits is added to its reversal, the result is a multiple of 11.

You can illustrate this amazing fact (Gardner 1979) with this example: $4798 + 8974 = 13772 = 11 \times 1252$. Prove this by representing the sum of a four-digit number and its reversal by $(1000a + 100b + 10c + d) + (1000d + 100c + 10b + a)$, combining like terms, and factoring 11 out of the result.

There are many tricks for mental computation in Cutler and McShane (1960). Among them is amazing fact #7:

Amazing Fact #7

To multiply a whole number n by 12, obtain each digit of the product by the rule: Double the corresponding digit of n, beginning with the units, and add it to the digit of n to its right; write down the units digit of the sum thus obtained, carrying when necessary.

We illustrate 12 × 597: Double 7, getting 14, and since there is no digit to the right of 7 in 597, simply write down the 4 and mentally carry the 1. The units digit of the product will be 4. To get the tens digit of the product, double the tens digit, 9, of 597, getting 18, add the digit to the right of 9, namely 7, and add the 1 that was carried, giving 26. Write 6 as the tens digit of the product, and carry the 2. Proceed to the hundreds digit by doubling 5 (the hundreds digit of 597), adding the 9 on its right, and adding the 2 that was carried, getting 21. Write the 1 in 21 as the hundreds digit of the product, and carry the 2. Since there is no (thousands) digit to the left of 5 in 597 to double, simply add the 2 that was carried to the 5, obtaining 7 as the thousands digit of the product. The product is 7164.

The algebra used to prove this fact is far simpler to explain than the mechanics of the trick itself. Represent the product of 12 and a three-digit number as $(10 + 2)(100a + 10b + c)$, expand using polynomial multiplication, and rewrite the result in the form

$$1000a + 100(2a + b) + 10(2b + c) + 2c.$$

Using algebra in presentations such as the ones described here helps students realize that they are learning algebra "for mathematics' sake" as well as for the many uses they will encounter in real-world applications. Furthermore, they have fun learning these examples.

REFERENCES

Cutler, Ann, and Rudolph McShane, trans. *Trachtenberg Speed System of Basic Mathematics.* Garden City, N.Y.: Doubleday & Co., 1960.

Demir, Huseyin. "Quickies." *Mathematics Magazine* 59 (April 1986): 112, 120.

Fawcett, Harold P., and Kenneth B. Cummins. *The Teaching of Mathematics from Counting to Calculus.* Columbus, Ohio: Charles E. Merrill Publishing Co., 1970.

Gardner, Martin. *Mathematical Circus.* New York: Alfred A. Knopf, 1979.

Jacobs, Harold R. *Mathematics: A Human Endeavor.* San Francisco: W. H. Freeman & Co., 1970.

29

Teaching Absolute Value Spirally

Alex Friedlander
Nurit Hadas

$\mathbf{S}$ OLVING inequalities that involve absolute values frequently turns out to be an unrewarding experience for both algebra teacher and algebra student. Most textbooks define the absolute value of a number as the distance of the corresponding point on the number line from the origin, but this graphical approach is quickly abandoned in favor of a numerical definition (i.e., the number itself if it is positive, its inverse if negative). Consider the algebraic solution of the inequality $|x + 2| < 4$ or, even worse, $|x + 2| < |x - 4|$. In the first example, the student is usually required to transform the original inequality to $-4 < x + 2 < 4$ and proceed algebraically. In the second, the system of inequalities obtained by removing the absolute value signs is beyond the capabilities of most students.

In a discussion among a group of teachers who taught this topic in first- and second-year algebra, we had begun by reviewing methods of teaching absolute value. We were surprised to find quite a few relevant papers that presented a variety of methods, ranging from the purely graphical to the purely algebraic.

Sink (1979) offered a systematic representation of the algebraic solutions of such open sentences (see fig. 29.1). His approach clearly outlines the different cases involved in absolute value sentences and the relationship among them. It seems, however, that even if one masters the mechanical manipulations of a "solution by cases"—and our experience with students and teachers suggests that this is unlikely—the meaning of the original problem and its solution are lost in the process.

Another way to solve inequalities was suggested by McLaurin (1985): If we transform the given inequality into an equation and then solve it, the "critical points" obtained will determine several intervals on the number line, with each of them either completely included or completely excluded from the solution set. Therefore, in order to find the solution set, it is sufficient to check one random number from each of these intervals. This approach has the advantage of offering a unified method for solving inequalities of different kinds, but it does not stress the concepts that are specifically related to the absolute value. As with any algebraic solution, this method is also bound to lead some students to a mechanical, and not always correct, solution of the inequality at hand (see also Dean [1985] and Roberti [1985]).

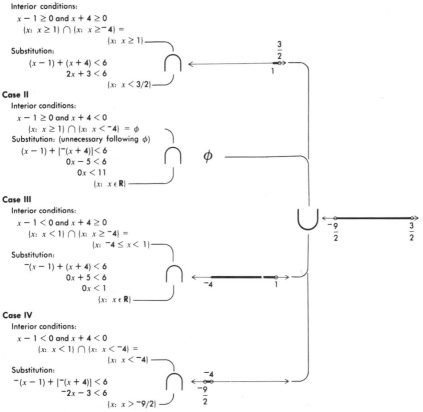

Fig. 29.1. Solution "by cases" of $|x - 1| + |x + 4| < 6$ (from Sink [1979])

Brumfiel's article (1980) offered a good overview of five approaches to the concept of absolute value. Two of them consider $|a - b|$ as the distance between the corresponding points on the number line, with $|a|$ being a particular case for $b = 0$. The other three definitions emphasized the numerical aspect of absolute value:

$$|a| = \begin{cases} a \text{ if } a \geq 0 \\ -a \text{ if } a < 0 \end{cases} \qquad |a| = \max(a, -a) \qquad \text{or} \qquad |a| = \sqrt{a^2}$$

After reviewing these papers with the teachers, we came to the conclusion that rather than using a single approach, we would try to create a spiral sequence of activities that would present the concept of absolute value from various aspects and employ the mathematical knowledge that students have at different grade levels. In the following, we describe typical exercises to illustrate the approach. These exercises require a visual and more meaningful approach than that used in common algebraic methods. We start our sequence with the notion of distance on the number line and later employ coordinate-system graphics in order to solve more complex absolute value sentences.

DISTANCE ON THE NUMBER LINE

In the initial stages of teaching algebra, we focus on the set of real numbers and their representation on the number line, on finding solution sets for open sentences, and on translating word problems into algebraic models. All these topics are nicely combined in the number-line solution of absolute value sentences. At the prealgebra level, we may ask students to find the set of numbers situated at a distance less (or more) than a given distance from a given number.

Exercise 1

Use the number line to find the numbers whose distance from 1 is less than 2.

$$\begin{array}{c} \longleftarrow \!\!+\!\!+\!\!+\!\!+\!\!+\!\!+\!\!+\!\!+\!\!+\!\!+\!\!+\!\!\longrightarrow \\ \text{-4 -3 -2 -1 \; 0 \; 1 \; 2 \; 3 \; 4 \; 5} \end{array}$$

Later, when open phrases are introduced, we can reinforce the notion that $|a - b|$ represents distance by asking students to make the necessary connections given in exercise 2.

Exercise 2

Complete the following table:

Verbal Description	Open Phrase
The distance between a and 2 The distance between 2 and a	--------------------
-------------------- --------------------	$|a - 3|$ $|a + 3|$
The distance between a and b The distance between b and a	-------------------- --------------------

The classroom teaching of these exercises was not very exciting, but we found that they were a necessary preparation for the next stage—solving open sentences.

Exercise 3

Given $|x| < 3$,

(a) find several positive and negative numbers that belong to the solution set;

(b) mark on the number line the complete solution set.

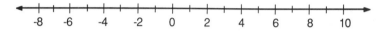

Open sentences, such as $|x - 1| = 2$, or $|x + 2| > 1$, can now also be solved without the use of the algebraic "case" method. Our experience in the classroom showed that the best results were achieved when the solution followed three stages:

Describe in words → Find solution set → Write solution set
(use notation of   (use number line)   (use algebraic
  distance)   notation)

The table in exercise 4 was designed to help students organize their solution.

Exercise 4

Complete the following table:

	Open Sentence	Verbal Description	Graph	Solution Set
(a)	$|x - 2| < 3$	Numbers whose distance --------	-3 -2 -1 0 1 2 3 4 5 6	$\{x\mid \ \}$
(b)	$|x + 1| > 2$		-5 -4 -3 -2 -1 0 1 2 3 4	$\{x\mid \ \}$
(c)	$\mid \quad \mid <$		-3 -2 -1 0 1 2 3 4	$\{x\mid \ \}$

In the classroom, many exercises of types (a) and (b) were presented before we required the student to find the inequality that corresponds to a given solution set, type (c). The latter problem caused difficulties, and students tended to relate their verbal description of the given segment to its extremes (i.e., "numbers between −1 and 3") rather than to distance from its midpoint (i.e., "numbers whose distance from 1 is less than 2"). A review of the previous examples helped us to generalize the structure of the verbal description of an absolute value inequality.

The three-stage method has many advantages:

• Translating into words and distance considerations on the number line encourages the student to keep the whole picture in mind rather than sink into mechanical manipulations of algebraic sentences.

• Translation from the mathematical form into words is a much-needed ability that is frequently neglected in algebra classes. Usually, we focus on word problems that require translation into algebra and not vice versa. Sometimes we do ask students to "write a story" that is based on a given algebraic sentence, but it seems to us that here the need to describe an algebraic sentence in words is given a more natural context. For instance, the solution of the first inequality in the table in exercise 4 is almost completed once the sentence "The numbers whose distance from 2 is less than 3 units" is written in the second column.

• In the classroom, we found that the search on the number line for numbers that satisfy a certain condition poses a challenge like that in a "detective story," and we frequently observed that students teamed up to make sure that no number "escaped" from the solution set. This remark is particularly valid for more complex sentences, such as the ones presented in exercise 5.

Exercise 5

Solve:

(a) $|x - 2| > |x - 6|$
(b) $|x - 3| + |x + 1| = 8$
(c) $|x - 3| - |x + 1| > 4$

In each of these problems, the numbers involved in the distance considerations divide the number line into three regions: to the left of the smaller, to the right of the larger, and between the two numbers. In our solution, we have to examine numbers from each of these regions.

Thus, problem (a) may be described verbally as "the numbers whose distance from 2 is greater than their distance from 6." Since the midpoint of the (2,6) segment is equidistant from 2 and 6, we conclude that the numbers to the right of 4 form the solution set.

Closer to 2 — Closer to 6

-2 -1 0 1 2 3 4 5 6 7 8

(a) $|x - 2| > |x - 6|$

The distance between the two points given in problem (b) is 4, and therefore the sum of the distance becomes 8 when we move 2 more units beyond either extremity of the $(-1,3)$ segment. These considerations allow us to conclude that -3 and 5 are the required solutions.

In problem (c), we are looking for "numbers that are more than 4 units farther from -1 than from 3." Since the distance of 3 from -1 is itself 4 units, any number to the right of 3 belongs to the solution set.

Ahuja (1976) suggested that distance and ratio considerations be combined to solve even more complex sentences such as $|x - 1| < |2x - 14|$. Complex absolute value sentences are more readily solved by using two-dimensional graphs, and their solutions should therefore be delayed until the topic of graphing functions is reached in algebra classes.

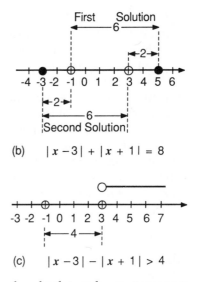

(b) $|x - 3| + |x + 1| = 8$

(c) $|x - 3| - |x + 1| > 4$

COORDINATE SYSTEM GRAPHICS

When the stage of graphing functions is reached, the definition of the absolute value as $|x| = \begin{cases} x \text{ if } x \geq 0 \\ -x \text{ if } x < 0 \end{cases}$ can be introduced as a graphing assignment. The exercises in graphing piecewise-defined functions that we give students are often quite arbitrary. The absolute value function offers a good opportunity to show a simple and natural example of such a function. Graphing the absolute value of any function (i.e., $|f(x)|$) is the next step in our sequence.

Exercise 6

Given $f(x) = x^2 - 2x$, graph $|f(x)|$.

For this particular problem, $f(x)$ is negative between its two zeroes (2 and 0), and the graph of $|f(x)| = |x^2 - 2x|$ is obtained from that

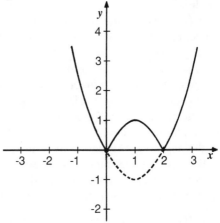

of $f(x)$ by reflecting the corresponding points of $f(x)$ in the x-axis.

After several similar examples, students may conclude that, as with $|x|$, $|f(x)|$ is a piecewise-defined function. This means that $|f(x)| = \begin{cases} f(x) \text{ if } f(x) \geq 0 \\ -f(x) \text{ if } f(x) < 0 \end{cases}$ and, more important for our needs, that the "negative branch" of the $f(x)$ graph is an x-axis reflection of the corresponding part of the graph of $|f(x)|$.

Once we know how to graph absolute value functions, we can expand our number-line approach to the graphical solution of absolute value sentences in the two-dimensional coordinate system (see also Arcidiacono 1983). The spiral approach allows us to present here exercises that were encountered before *and* add new, more complex ones.

Exercise 7

Solve:

(*a*) $|2x - 3| < 5$

(*b*) $|x - 2| < |x - 6|$

(*c*) $|2x - 1| < |x - 2|$

(*d*) $|x - 3| + |x + 1| = 8$

(*e*) $|x^2 - 5x + 6| < 1$

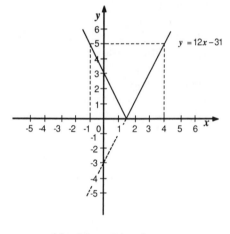

(a) $|2x - 3| < 5$

Following the spiral approach, we deliberately chose to use the new method on exercises that were posed before on the number line (*a, b, c, d*) and added some more complex examples that could not have been solved earlier (*e*). The solution of inequalities (*a*), (*b*), and (*d*) will serve as examples of this method.

In (*a*), we draw the graph of $y = |2x - 3|$ (not a difficult task, if it is done by reflecting the negative branch of $y = 2x - 3$) and then conclude that the function gets values less than 5 for $-1 < x < 4$.

In (*b*), we draw the graphs of $y = |x - 2|$ and of $y = |x - 6|$ and then conclude that the first function gets smaller values than the second for $x < 4$.

In order to solve (*d*), it is desirable to rearrange it as $|x - 3| - 8 = -|x + 1|$. The corresponding graphs and solutions (-3 and 5) can be obtained by the x-axis reflection of $y = |x + 1|$ and the vertical translation of $y = |x - 3|$ in the negative y-direction by 8 units.

We have seen in the previous section that open sentences (*b*) and (*d*) can be solved by using distance considerations on the number line. The coordinate system, however, has some advantages:

• It allows us to solve absolute value inequalities of greater complexity. If the inequality contains more than one absolute value (as in inequalities b, c, and d) or if the coefficient or the exponent of the variable included in the absolute value is other than 1 (as in a, c, and e), a solution on the number line becomes difficult or impossible.

• It employs a more or less uniform strategy for all the cases. Further, it can be used later to solve inequalities of any kind.

• It is more accessible to students of average and below-average ability. As we saw in the previous section, solving inequalities (b) and (d) on the number line requires quite refined mathematical thinking from beginning algebra students.

• It presents one of the few cases in the mathematics curriculum in which

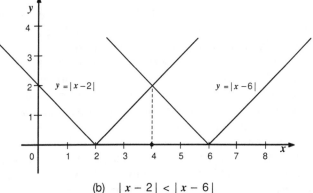

(b) $|x - 2| < |x - 6|$

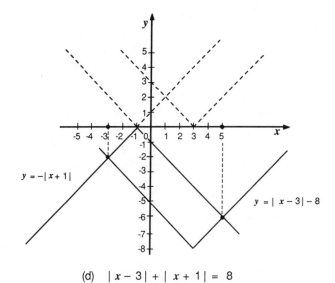

(d) $|x - 3| + |x + 1| = 8$

a graphical solution is less tedious and time-consuming than the algebraic solution. We frequently require students to solve linear and quadratic equations graphically, or even worse, graphically *and* algebraically, whereas both the teacher and the student know that the algebraic approach is, in most of these instances, more convenient.

• It employs the graphical skills of symmetry, reflection, and translation in a meaningful way.

SUMMARY

The dissatisfaction of both teachers and students with the mechanical and technical difficulties involved in an algebraic solution of absolute value sentences led us to a sequence of related activities that allow a meaningful approach to the topic.

Many suggestions have been made lately with regard to teaching the concept of absolute value. In this paper we considered these suggestions and attempted to sequence them into a spiral approach that allowed the teaching of this topic at various stages of the algebra curriculum. At each stage, the sequence expands the student's ability to solve absolute value sentences of increasing complexity. Thus, the student is led through a small-scale model of mathematical development. At each stage, open sentences that were previously difficult or impossible to solve now become accessible. The gradual development also has the advantage of reinforcing the topic.

Reactions from teachers who taught absolute value sentences according to the sequence described above were definitely positive. In comparison to the regular approach, student motivation, understanding, and achievement were higher.

Our experience indicates that for the first years of algebra the one- and ·two-dimensional graphical approach has many advantages over the algebraic "solution by cases" of absolute value sentences. The latter method should, in our view, be left to higher-ability, advanced algebra students or abandoned altogether.

REFERENCES

Ahuja, Mangho. "An Approach to Absolute Value Problems." *Mathematics Teacher* 69 (November 1976): 594–96.

Arcidiacono, Michael J. "A Visual Approach to Absolute Value." *Mathematics Teacher* 76 (March 1983): 197–201.

Brumfiel, Charles. "Teaching the Absolute Value Function." *Mathematics Teacher* 73 (January 1980): 24–30.

Dean, Eleanor. "Reader Reflections: Inequalities I." *Mathematics Teacher* 78 (November 1985): 590.

McLaurin, Sandra C. "A Unified Way to Teach the Solution of Inequalities." *Mathematics Teacher* 78 (February 1985): 91–94.

Roberti, Joseph V. "Reader Reflections: Inequalities II." *Mathematics Teacher* 78 (November 1985): 590, 651.

Sink, Stephen C. "Understanding Absolute Value." *Mathematics Teacher* 72 (March 1979): 191–95.

30

Which One Doesn't Belong?

David J. Glatzer

WHICH of the equations below doesn't belong?

$$3(x + 4) = 3x + 12$$
$$3(x - 7) = 3x - 21$$
$$5(x + 4) = 5x + 9$$
$$5(x - 2) = 5x - 10$$

When asked which of these equations "doesn't belong," algebra and prealgebra students are likely to say that the third equation is the oddball because it represents an incorrect application of the distributive property whereas the others show a correct application of the property.

This article suggests using the "Which One Doesn't Belong?" activity in teaching algebra when a review or reinforcement of concepts and skills is appropriate. Typically, the teacher might offer five or six examples on transparencies or on card stock as a warm-up or closing activity. The directions should normally be open-ended, namely, "Which one doesn't belong?" However, there may be times when the directions should be more focused. Here is an example of directions that focus attention to one area:

Three of the following equations represent lines passing through the origin. One does not pass through the origin. Which is it? Explain your answer.

$$y = 3x \qquad y = 2x + 5 \qquad y = -4x \qquad y = x$$

There are advantages to the open-ended directions. Students will spend time and effort coming up with arguments supporting different choices for the oddball. When the argument is valid, the choice is correct. Thus, there is no single "correct" answer.

Instructional procedures that enhance the effectiveness of this activity need to be used. One approach would be to have students make a written commitment to an option and then to write a rationale. At this stage, the student might compare the selection and rationale with those of a neighbor. These discussions might precede a discussion by the entire class.

Below are fifteen examples from the algebra curriculum. Add additional examples of your own design.

1.	$6x = 18$	$6x = 22$	$6x = 42$	$6x = 36$
2.	$\sqrt{36}$	$\sqrt{81}$	$\sqrt{100}$	$\sqrt{200}$
3.	1.4	$\sqrt{2}$	$\sqrt[4]{4}$	$2^{\frac{1}{2}}$
4.	$x + y = 5$	$y = x^2$	$x^2 + y^2 = 9$	$y = x^3$
5.	1	8	1000	81
6.	$4x^2 - 9$	$16a^2 - 25$	$25b^2 - 1$	$49c^2 - 8$
7.	associative	repetitive	commutative	distributive
8.	32	1024	25	128
9.	$x^2 + 6x + 9$	$x^2 - 2x + 1$	$x^2 + 20x + 100$	$x^2 + 5x + 6$
10.	i^{50}	i^{48}	i^{36}	i^{44}
11.	$\log_2 2$	7^0	$\log_3 9$	$\log_5 5$
12.	$x^2 = 9$	$x^3 = 27$	$\mid x \mid = 3$	$x = \pm 3$
13.	$5x^3y$	$6x^3y^2$	x^3y^2	$-6x^3y^2$
14.	$1,7,13,19,\ldots$	$1,4,16,64,\ldots$	$4,7,10,13,\ldots$	$22,33,44,55,\ldots$
15.	$(-2)^5$	$-3^1 - 1$	$2^5 - 2^6$	$\mid -2^5 \mid$

Although the examples above involve algebraic concepts and skills, geometric or arithmetic topics could just as well be used. In fact, topics could be chosen from other subject areas or from nonschool areas of interest to students, such as science, music, geography, sports, and rock music.

Another modification has the students generating the items in the examples. A typical assignment would have students responsible for creating a "baker's dozen" examples: six mathematical, six nonmathematical (based on areas of interest), and one "stumper." Another challenging use of the format would be for students or the teacher to generate one example where each option could be the oddball based on some specific rationale, for example,

$$1 \qquad 2 \qquad 7 \qquad 13.$$

In this example, "1" is not a prime number; "2" is not odd; "7" is not a Fibonacci number; and "13" is not a one-digit number.

In summary, "Which One Doesn't Belong?" is an activity offering flexibility within the context of review and reinforcement. Skills and concepts can be highlighted as divergent thinking is promoted. Try your own modifications of the format!

31

Integrating Statistical Applications in the Learning of Algebra through Problem Solving

Carolyn A. Maher
John P. Pace
John Pancari

S CHOOL algebra, a highly refined collection of ideas, may appear to new learners to be a remote, formal system with little application to their daily lives. When ideas are presented in a problem that does have direct application to the learner, motivation is more likely. Moreover, when the ideas are presented in a setting in which the learner is active in constructing the knowledge, sustained learning is likely to take place. Problems can be an effective motivational device when they are real and of concern to the learner.

The use of real data to promote the learning of statistics has been widely proposed (Landwehr and Watkins 1986; Maher and Pace 1985; and Maher 1981). Recently, the Joint Committee of the American Statistical Association and the National Council of Teachers of Mathematics developed units endorsing the use of real data in mathematics courses as well as in probability and statistics courses in schools (Landwehr et al. 1986). This article describes such an activity used with a high school algebra class. The example deals with the analysis of the performance of the school's baseball team. Sports data such as these, usually collected by team managers and student assistants, provide a rich source for generating problems that can stimulate the learning of relations, functions, and their graphic representations. A sports problem formulation, with teacher guidance, can include the identification of variables that might contribute to a team's success and eventually to the formulation of hypothesized relationships among these variables. The

Special thanks are owed to the participating students of St. Joseph High School of Hammonton, New Jersey.

223

subsequent test of these hypotheses can then employ x,y graphs, functions, and hypothesis-testing procedures.

During the course of dialogue between students and teacher, the following variables are likely to be mentioned as important in determining the quality of a baseball team's performance: times at bat, runs scored, strikeouts, stolen bases, singles, doubles, triples, home runs, errors, hits, walks, and runs batted in. After identifying the potentially relevant variables, students can then be asked if any of the variables are pairwise related. Many pairs are likely to be suggested, such as number of times at bat and number of hits. Ask students to select possible factors that influence the performance of their school's baseball team. The following pairs of variables were offered by one algebra class: strikeouts and walks; hits and runs batted in; extra-base hits and strikeouts; singles and stolen bases; extra-base hits and runs batted in.

Now, either the teacher or the students can direct particular attention to one suggested pair of variables. The ensuing discussion of the relationship between these two variables can include the contruction of a scatter plot.

A simple graphical method (Tukey 1977) that makes use of the scatter plot can explore bivariate relationships through the construction of a median trace. For the scatter plot of the number of times at bat versus the number of hits, we proceed to construct the median trace graph (fig. 31.1) by the following steps:

1. Draw a vertical line through the median x value on the scatter plot to divide the data into two sections. (The median of a set of numbers is (a) the middle data value if the number of data values is odd and (b) the average of the *two* values if their number is even.)

2. Draw a vertical line through the median x value of each half of the data to form four sections. (Tukey's procedure suggests that the splitting process be continued for the outermost sections whenever nine or more data values remain in these sections. When one or more values falls on a section boundary line, include the boundary data values when determining the x or y median of that section.)

3. Find, for each section, the intersection of its median x and median y lines. (These intersection points are called section medians.)

4. Connect the section median points consecutively with broken line segments to form the median trace.

Students can be asked to compare the relationship described by the median trace to other functions. For example, one might ask if it is reasonable to expect a player to get a hit for each time at bat ($y = x$) or to expect the number of hits to be independent of the number of times at bat ($y = 22$, where 22 is the median hit value). These functions can be seen in figure 31.2.

An examination of figure 31.2 suggests that neither $y = x$ nor $y = 22$ are

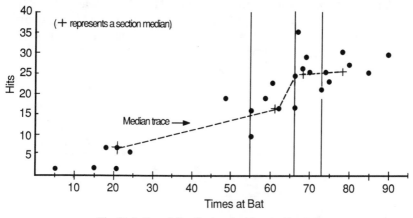

Fig. 31.1. Examining the trend with a median trace

reasonably close to the median trace. The class in question proposed that a better relationship might be one in which the number of hits is proportional to the number of times at bat ($x = mx$) and where m, the constant of proportionality, is the team batting average calculated by dividing total hits by total times at bat ($m = 463/1397 = 0.33$). The result of graphing $y = .33x$ is also shown in figure 31.2.

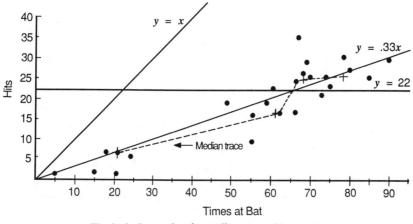

Fig. 31.2. Comparing the median trace with $y = .33x$

Students can now discuss the fit of the function $y = .33x$ as well as others that might be suggested.

As an indication of the variability of the data, upper and lower quartile traces can be constructed in a manner analogous to the median trace. As in the construction of the median value, 25th and 75th percentile estimates are constructed within each section of the data. This procedure will generate a region on the scatter plot containing approximately the middle 50 percent of the data points and is illustrated in figure 31.3. Note that $y = .33x$ is contained within the middle region formed by the upper and lower quartile traces, whereas $y = x$ and $y = 22$ are not. This suggests that $y = .33x$ provides a more reasonable fit to the data.

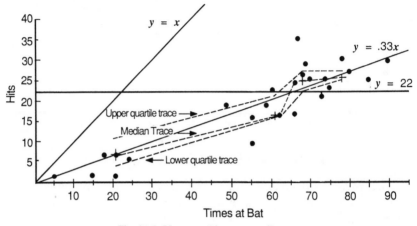

Fig. 31.3. Upper and lower quartile traces

The class, organized in groups, chose to examine other relationships that might characterize the team's performance. Figure 31.4 shows the relationship between hits and runs batted in with $y = .6x$, where the slope value of .6 was derived by dividing total runs batted in by total hits.

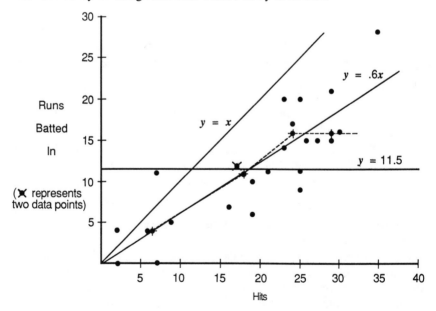

Fig. 31.4. Hits versus runs batted in

Figure 31.5 illustrates the relationship between extra-base hits and runs batted in with $y = 2.4x$. The slope value of 2.4 was derived by dividing total runs batted in by total extra-base hits.

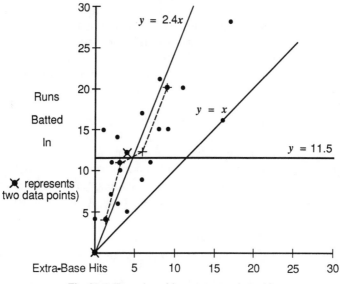

Fig. 31.5. Extra-base hits versus runs batted in

Compare figures 31.4 and 31.5 and note in the latter the much stronger relationship of runs batted in to extra-base hits.

Finally, figure 31.6 depicts an example of extra-base hits versus strikeouts. Here, students found no clear linear relationship.

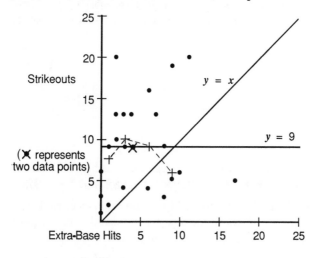

Fig. 31.6. Extra-base hits versus strikeouts

CONCLUSION

Students are more likely to be motivated to learn a subject when class-room problems are real and when they identify and help select the problems for exploration. They are more likely to find learning meaningful when the conceptual knowledge in a discipline arises in response to the need to answer questions posed by these problems. We are suggesting that mathematics, generally, and algbebra, particularly, be viewed as a problem-solving activity. The students should be allowed to ask questions, assist in defining problems, suggest paths of solution, and help interpret the results. The teacher should guide the process in an organized way toward goals that represent the content of the subject—in this instance, algebra.

There will undoubtedly be variations in individual students' reconstruction of the subject for themselves. Classrooms in which teachers assess student interest and their level of understanding through active dialogue and where problem-solving activities serve as the context for motivating students to learn new topics are more likely to promote student construction of content.

REFERENCES

Landwehr, James, and Ann Watkins. *Exploring Data.* Palo Alto, Calif.: Dale Seymour Publications, 1986.

Landwehr, James, Ann Watkins, Claire Newman, Thomas Obremski, Richard Scheaffer, Mrudulla Gnanadesikan, and Jim Swift. *Quantitative Literacy Series.* Palo Alto, Calif.: Dale Seymour Publications, 1986.

Maher, Carolyn Alexander. "Simple Graphical Techniques for Examining Data Generated by Classroom Activities." In *Teaching Statistics and Probability,* 1981 Yearbook of the National Council of Teachers of Mathematics. Reston, Va.: The Council, 1981.

Maher, Carolyn, and John Pace. "Success with Statistics through Planned Experiments." *Mathematics and Computer Education* 19 (Spring 1985): 93–98.

Tukey, John W. *Exploratory Data Analysis.* New York: Addison-Wesley Publishing Co., 1977.

CAN YOUR ALGEBRA CLASS SOLVE THIS?

Problem 22. The township population increased from x thousand to $x + y$ thousand. Find the percent of population increase.

Solution on page 248

32

Input-Output Modifications to Basic Graphs: A Method of Graphing Functions

John S. Thaeler

THE graph of a function can be drawn by the tedious process of calculating and plotting points. Input-output graphing, however, is far easier. This method recognizes that certain modifications affect all graphs in the same way without regard for the initial shape of the basic graph. For example, multiplying a basic function by 2 will cause its graph to be stretched by a factor of 2 in the y direction, away from the x-axis. Learning what these modifications are and how each affects the graph of any function can greatly simplify the process of graphing. Input-output graphing is described in the following steps.

Basic Functions

The familiar graphs shown in figure 32.1 are fundamental to using input-output graphing techniques. The title given with each graph tells how that function will be identified in the black box described in the next step.

Filling in the Black (Function) Box with the Rule of the Function

Before graphing a function, solve its equation for y so that the function is expressed with the input variable, x, appearing only once on the right side. For example, the function $y = -2x^2 + 12x - 14$ will be $y = -2(x - 3)^2 + 4$ after completing the square. Now, starting with the x, describe in words the step-by-step process of what happens to the x to turn it into the output y. For the function given above, the black box is shown in figure 32.2.

Input or Output?

Decide which modifications to the basic function are input modifications and which are output modifications. First circle the basic function in the

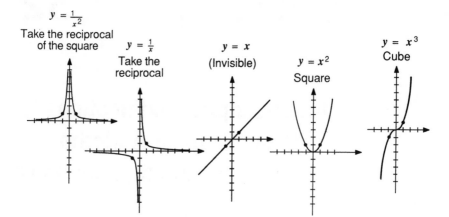

$y = \dfrac{1}{x^2}$
Take the reciprocal of the square

$y = \dfrac{1}{x}$
Take the reciprocal

$y = x$
(Invisible)

$y = x^2$
Square

$y = x^3$
Cube

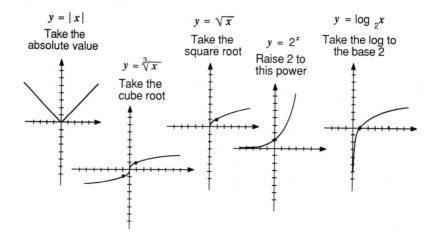

$y = |x|$
Take the absolute value

$y = \sqrt[3]{x}$
Take the cube root

$y = \sqrt{x}$
Take the square root

$y = 2^x$
Raise 2 to this power

$y = \log_2 x$
Take the log to the base 2

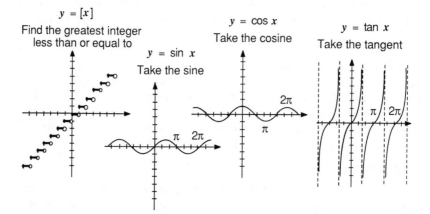

$y = [x]$
Find the greatest integer less than or equal to

$y = \sin x$
Take the sine

$y = \cos x$
Take the cosine

$y = \tan x$
Take the tangent

Fig. 32.1

black box. If you have trouble deciding which entry in the box is the basic
function, refer to the titles of the
graphs in figure 32.1. The basic function determines the basic shape of
the graph. In our example, the basic
function is SQUARE, and so we know
that the graph will be some modification of the parabola.

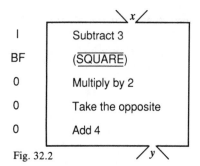

I	Subtract 3
BF	(SQUARE)
0	Multiply by 2
0	Take the opposite
0	Add 4

Fig. 32.2

• *Any modifications (entries in the black box) occurring before, or above,
the basic function are input modifications.* They are closer to the top of the
box. Such modifications, with examples shown below, will affect the basic
graph in the *x* (input), or left-right, direction:

Stretch or compress the graph in the *x* direction either toward or away
from the *y*-axis.

Flip the graph left/right.

Shift the graph left or right.

• *Any modifications occurring after, or below, the basic function will be
output modifications.* These changes are closer to the bottom of the box.
Such modifications (examples below) will affect the graph in the *y* (output),
or up-down, direction:

Stretch or compress the graph in the *y* direction either toward or away
from the *x*-axis.

Raise or lower the graph.

Flip the graph up or down.

Obviously, it is vital to be able to identify the basic function with confidence. Once it is identified, input and output modifications can be determined by their position relative to the encircled basic function.

Specific Output Modifications

We shall describe output modifications first because they are simpler and
more direct in their effect on the graph of the basic function. Remember that
their unifying theme is that they *act on the curve in the y, or up and down,
direction.*

Multiplying by a positive number, a. If *a* is greater than 1, the graph of the
basic function is stretched away from the *x*-axis by a factor of *a* in the *y*
direction. If *a* is between 0 and 1, the graph is compressed vertically toward
the *x*-axis. This stretching and compressing can be done with a reasonable
degree of accuracy by drawing a series of vertical lines through the graph.
Then, for example, if *a* has a value of 2, double the distance from the *x*-axis to

the basic curve to locate a point on the new (stretched) graph (see fig. 32.3).

If a has a value of $1/3$, chop or compress the distance between the x-axis and the original graph to $1/3$ of its original size to locate the new (compressed) curve (see fig. 32.4).

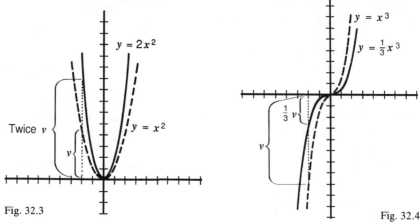

Fig. 32.3

Fig. 32.4

The justification for this modification is that each output from the basic function is multiplied by a, and thus the vertical distance from the x-axis to the curve is stretched or compressed by this factor, a.

Taking the opposite. As an output modification, taking the opposite has the effect of flipping the graph in the y direction (up to down, or down to up) using the x-axis as the flip line, or line of reflection. See figure 32.5.

The justification for this modification is that every output has its sign reversed by taking the opposite. Thus, if a point on the basic curve was plotted by going to the right and *up*, now a point with the same value of x would be plotted by going to the right and *down*. Overall, the effect is to flip the whole original graph to produce the new graph. The portion of the original graph that was above the x-axis produces the part of the new graph that is below the x-axis; the portion of the original graph that was below the x-axis produces the part of the new graph that is now above the x-axis.

Adding a positive or negative number. When some number, c, is added as an output modification, the effect on the basic graph is to raise it (when c is greater than 0) or lower it (when c is less than 0) by the absolute value of c units. For exam-

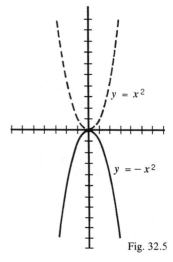

Fig. 32.5

ple, if $c = 2$, then the graph is raised two units. If $c = -3$, then the curve is lowered three units. See figure 32.6.

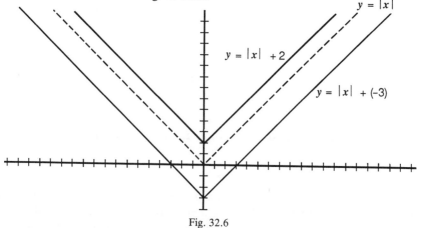

Fig. 32.6

The justification for this modification is that adding the value of c to every output increases (when c is greater than 0) or decreases (when c is less than 0) the distance from the x-axis to the curve everywhere by exactly the amount $|c|$.

Specific Input Modifications

Input modifications have the unifying theme of acting on the graph in the x (left or right) direction. Because they act before the basic function, they have an almost inverse effect on the graph and thus require special formats to allow easy graphing.

Subtracting a positive or negative number. As an input modification, adding or subtracting a number has the effect of moving the graph either left or right. To allow the usual meaning of positive (shift right) and negative (shift left) to apply, this input modification must be rewritten in the format "$x - $ something." Then if the "something" is positive, shift the graph to the right that many units. If the "something" is negative, shift the graph that many units to the left. The justification for this special format can best be explained with an example. The parabola $y = x^2$ has its minimum output of 0 when the function is squaring (has an input of) 0. By contrast, $y = (x - 3)^2$ will also have its minimum, or lowest, point graphically when the function is squaring 0. But this will now occur when the input x has a value of 3. Thus the minimum point of the parabola is now located at the point (3,0). That is, the graph has been shifted to the right three units. For the function $y = (x + 2)^2$ the parabola will have its minimum at $(-2, 0)$, since the function will be squaring 0 when $x = -2$. By writing $x + 2$ as $x - (-2)$—that is, in the format $x - $ something—we find that the correct motion of shifting the graph two units to the left is apparent. See figure 32.7.

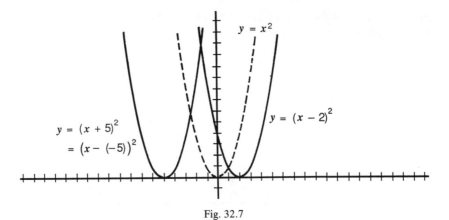

Fig. 32.7

Multiplying or dividing by something. This input modification also requires a special format to make the expected rules for stretching and compressing hold. (We say *expected* in the sense that multiplying by something larger than 1 should stretch the graph, and multiplying by something between 0 and 1 should compress it.) The format needed to make this expectation work is "x over something," or

$$\frac{x}{\text{something}}.$$

Now when the something is greater than 1, the graph is stretched in the x direction away from some vertical line, often the y-axis. If the something is between 0 and 1, the graph is compressed toward the vertical line. To actually do the stretching or compressing accurately, draw horizontal lines on the basic graph and then double, triple, or halve the distance from the vertical line, such as the y-axis, to the graph. For example, $y = (2x)^2$ would be graphed by first rewriting it in the format

$$y = \left(\frac{x}{\frac{1}{2}}\right)^2,$$

showing that the something is $1/2$. Thus the basic curve must be compressed toward the y-axis so that the horizontal distances from the y-axis to the curve are half what they were for the basic function. See figure 32.8.

To graph the function $y = (x/3)^2$, we would note that this function is already in the proper format and that the something is 3. Thus we would need to triple the horizontal distance from the y-axis to the original graph

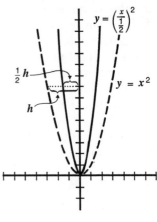

Fig. 32.8

to sketch the graph of this function (fig. 32.9).

The justification for this last modification can be understood through these examples:

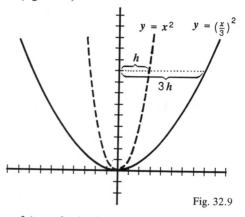

Fig. 32.9

- The standard parabola has an output of 1 when 1 is being squared. The parabola $y = (2x)^2$ will also have an output of 1 when 1 is being squared, but this occurs when x has a value of $1/2$. Thus an input of $1/2$ on the modified curve has the same output (namely 1) as an input of 1 on the basic curve. So the effect of this modification is to cut in half the horizontal distance to the curve from the y-axis. Putting the function into the special format "x over something" makes this halving of the distance apparent just by looking at the value of the something.

- For $y = (x/3)^2$, the modified curve now requires an input of 3 to actually be squaring a 1 (and thus have 1 for an output). This means that the horizontal distance to the basic curve from the y-axis must be tripled to produce the modified curve. Again the use of the special format makes this tripling effect apparent, since the something is a 3 when the function is written in the proper format, "x over something."

Taking the opposite. As an input modification, taking the opposite again has the effect of flipping the graph, but this time the flipping is done in the x direction (left to right, or right to left) using a vertical line (often the y-axis) as the flip line. In figure 32.10, the graph of $y = \sqrt{-x}$ is derived from the graph of the basic function $y = \sqrt{x}$ by flipping this graph from right to left using the y-axis as the flip line.

The justification for this type of modification is that inputs undergo a change in sign before they are processed by the basic function. Thus if the

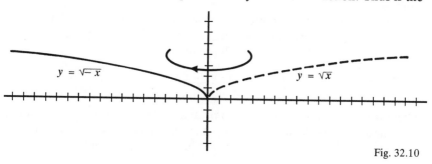

Fig. 32.10

point (p, q) lies on the basic graph, then the point $(-p, q)$ will have to appear on the modified graph. But these two points are symmetrical with respect to the y-axis. This means that one point will lie on top of the other if either is reflected in the y-axis.

Changing Input Modifications into Output Modifications

Because input modifications require a special format, they are often more difficult to graph. Some input modifications can be changed into output modifications algebraically and then graphed more simply.

The function $y = (2x)^2$, which was graphed earlier as an input modification (fig. 32.8), can be drawn more easily by rewriting it as $y = 4x^2$. Now the modification should be viewed as an output involving stretching vertically by a factor of 4.

As another example, consider the graph of $y = (-x)^3$. To draw this graph, flip the standard cubic horizontally, using the y-axis as the flip line. The results, as shown in figure 32.11, would be indistinguishable from the result of flipping the basic cubic vertically. This is not surprising, since $y = (-x)^3$ can be rewritten as $y = -x^3$. The latter would suggest flipping the standard cubic vertically to perform the needed output modification of taking the opposite.

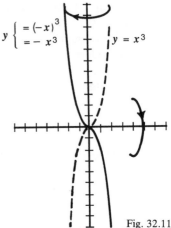

Fig. 32.11

Combining Several Modifications

When several modifications need to be done to create the graph of a function, a step-by-step process should be followed initially. First the basic graph is drawn, as represented by the circled entry in the function box. Then the input modifications are performed, followed by the output modifications.

Although the output modifications can probably be done in any order, input modifications are more difficult to combine. If shifting and stretching or compressing are both present, the required format is

$$\frac{x - \text{something}}{\text{something}},$$

where the somethings don't have to be alike. "Taking the opposite," when present, should be the last of the input modifications. For example, if $y = |3 - 2x|$ is to be graphed, it should first be rewritten in the form

$$y = \left| - \frac{x - \frac{3}{2}}{\frac{1}{2}} \right|.$$

To draw the graph, first write out the function box (fig. 32.12).

After starting with the V-shaped absolute value function (a), the basic graph is shifted to the right 3/2 units (b). Then it is compressed horizontally (c) by a factor of 1/2. This compressing is done with reference to the line $x = 3/2$. Finally it is flipped horizontally (d) in the line $x = 3/2$, which—because of the graph's symmetry—happens not to change it. These steps are shown in figure 32.13.

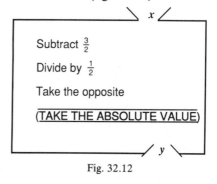

Fig. 32.12

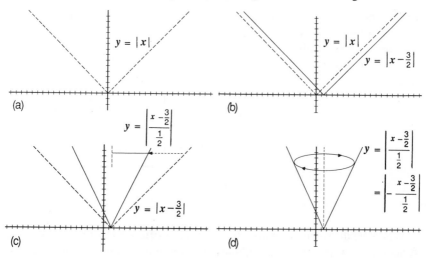

Fig. 32.13

As another example, let's draw the graph of $y = -2(x - 3)^2 + 4$, the function whose black box appears in figure 32.2. Figure 32.14 shows how the graph begins with a standard parabola (a). This is shifted to the right three units (b), since the something is 3. Next, the graph is stretched vertically (c) by a factor of 2, slipped upside down (d), and finally raised four units (e).

As a final example, let's apply the technique to drawing the graph of a sine function, $y = 2 + 3 \sin (2x + \pi/2)$. Use TAKE THE SINE as the basic function.

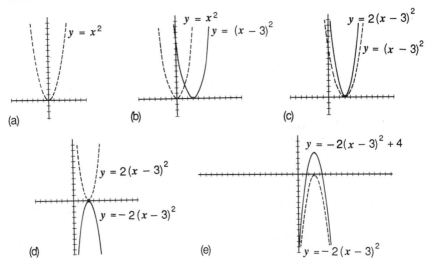

Fig. 32.14

Inside the parentheses the format can be improved by factoring out a 2, writing addition as subtraction of the opposite, and dividing by 1/2 instead of multiplying by 2. We have

$$y = 2 + 3 \sin\left(\frac{x - \left(-\frac{\pi}{4}\right)}{\frac{1}{2}}\right).$$

The function box appears in figure 32.15.

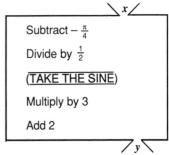

Fig. 32.15

When we draw trigonometric functions, it is often easier to draw the basic function and then put the coordinate system on the graph, labeling the axes to fit the curve. If needed, the graph can then be easily redrawn on a standard labeled system. So we shall start with a basic sine wave drawn on a horizontal line as shown in figure 32.16 (*a*). We can think of a sine wave as "starting" by rising from the *x*-axis. Thus if the first input modification involves shifting the graph to the left π/4 units (*b*), then the rising from the *x*-axis will occur at coordinate −(π/4). Dividing by 1/2 will cause the standard period of 2π to be compressed to only π units (*c*). Thus the graph will complete a cycle and start to rise from the *x*-axis again at 3/4π. They *y*-axis can be located by noting that a full cycle takes π units. We need to locate the axis one-quarter of a cycle to the right of the rising of the graph from the *x*-axis. Thus the

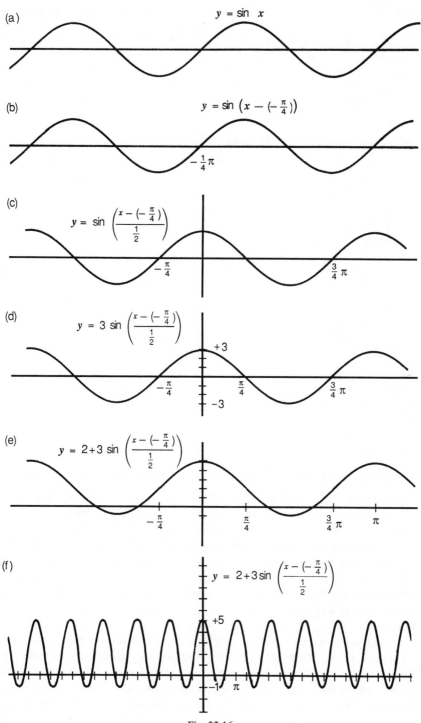

Fig. 32.16

y-axis goes through the peak (maximum) point of the curve. Multiplying by 3 means that the maximum and minimum points of the curve will now be $+3$ and -3 instead of $+1$ and -1 for the standard curve (d). Finally adding 2 will raise the whole curve two units (e). This means that the maximum point has a y-coordinate of $+5$ and the minimum a y-coordinate of -1 (f).

Limitations of the Input-Output Method of Graphing

In general, the input-output method cannot be applied if the input variable x occurs in more than one place in the rule of the function. An exception, as mentioned earlier, is when completing the square can be employed to express the rule, using x in just one place.

Even though x occurs only once in the function

$$g(x) = \frac{4}{x^2 + 4},$$

we still cannot use the input-output technique, because the function involves more than one basic function. Writing out the black box gives us figure 32.17.

The presence of the two basic functions, SQUARE and TAKE THE RECIPROCAL, means that we shall have to use some other technique to draw the graph of this function.

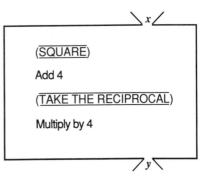

(SQUARE)

Add 4

(TAKE THE RECIPROCAL)

Multiply by 4

Fig. 32.17

Advantages and Disadvantages of the Input-Output Approach

One significant advantage of the input-output approach to graphing is that students learn patterns that apply to any graph for which a basic function is known. Having learned to graph parabolas and cubics, the student can apply the same approach to exponential, logarithmic, and trigonometric functions. Another important benefit is that they are able to determine the equation of a function, given its graph. The student asks what shifting, flipping, stretching, or compressing is necessary to produce the given graph starting with the basic function. Each motion contributes some quantity to the equation of the function.

For example, suppose you were given the graph shown in figure 32.18. This is obviously the graph of a parabola that has been shifted to the left three units, lowered one unit and turned upside down. Having the basic shape of a parabola means that the function box and equation will involve squaring. So we begin our equation construction with $y = x^2$. Shifting to the left three units is an input, or x direction, modification that can be accounted for by the format "x − something," where the something is (-3). Thus we have $y = (x - [-3])^2 = (x + 3)^2$. To turn the graph upside down we need to do an output modification that involves flipping. The step in the function box called "Take the opposite" accomplishes this flipping. Thus our shifted and flipped graph has the equation $y = -(x + 3)^2$. Finally we need to lower the graph one unit. Since lowering occurs in an output, or y, direction, "Subtract 1" will accomplish this step. We conclude that $y = -(x + 3)^2 - 1$ is the desired equation of our function. We check our results by noting that the y-intercept on the graph occurs at -10, and this agrees with the value of y from our equation when $x = 0$.

To check for stretching or compressing, one has to really know the shape of the basic curves. Then a careful comparison with the given curve will tell if any stretching or compressing has taken place. Suppose we want to know the equation of the graph shown in figure 32.19. We should recognize this graph as that of a square root that has been flipped to the left and moved to the left one unit. We can detect that the graph has also been compressed vertically by a factor of $1/2$ through noting that if we move one unit to the left of the vertex, we rise only a half unit to meet the curve instead of the usual one unit for a basic flipped square root. Or if we move four units to the left of the vertex, the graph climbs only one unit instead of the standard two units. Thus

our equation should be written as $y = \dfrac{1}{2}\sqrt{-(x - [-1])} = \dfrac{1}{2}\sqrt{-(x + 1)}$.

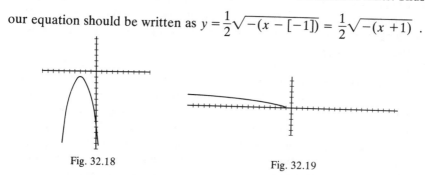

Fig. 32.18   Fig. 32.19

Similar comparisons of the given graph to the basic graph should reveal if any stretching or compressing has taken place. Even though such stretching or compressing could also be interpreted as horizontal compressing or stretching, it is easier to deal with this phenomenon, when possible, as an output modification. The main disadvantage of the approach is the need for an "x − something" or "x over something" format to make input modifications fit the mold. Students should practice the algebra steps needed to obtain the proper format.

Despite this limitation, however, the input-output approach gives students a valuable tool in understanding and gaining a feel for what the graph of a function is and how each number in the equation affects the final appearance of the graph.

33

Making Algebra Homework
More Effective

Gregory Holdan

MORE than one trial is usually required to master a skill in algebra. Consequently, practice of some sort is generally a planned part of algebra instruction. To ensure proper understanding of an algebraic skill or concept, an appropriate amount of guided practice should first be provided; however, at some point in the learning sequence the learner must be given the opportunity to engage in independent practice with the algebraic skill or concept. In mathematics classes, this usually means a homework assignment.

Research has generally shown that regularly assigned homework is preferable to no homework or to voluntary homework (Goldstein 1960; Coulter 1979; Rickards 1982). Several studies (Frederick and Walberg 1980; Keith 1982) have shown not only that the amount of time spent on homework contributes modestly, yet significantly, to improved grades but also that the effects of regularly completed homework may be cumulative (Goldstein 1960). There can be little doubt that homework is an integral part of instruction. Yet, of the many homework-related issues that might be considered, only a few have been investigated and shown to be significant.

SOME ELEMENTS OF EFFECTIVE PRACTICE

Research offers the following five principles for the algebra teacher to consider when designing homework.

1. Distributing practice over time is preferable to massing practice (Butcher 1975).

Traditional massed practice drills only the concepts and skills related to a single topic. Generally, a teacher teaches a topic, assigns practice on just that topic, discusses the assignment, and then begins teaching the next topic. Distributed practice, however, also includes practice with concepts and skills from previous lessons; in fact, proportionately more emphasis is on practice with previously learned concepts and skills than with new content (Dodes 1953).

Instead of assigning ten uniform motion problems all at once, distribute them so that students are required to do, for example, five problems one night, three the following night, and two the third night. Instead of assigning thirty trinomials to factor into the product of two binomials in one assignment, assign fifteen the first day, ten the next day, and five the following day. If exercises are distributed regularly, each assignment will include a natural mix of exercises representing your most recent two or three lessons. This results in homework assignments that structurally provide students with reinforcement over time, lessen the effects of forgetting due to the interference of new learning, and make the information more accessible when it needs to be remembered. Generally, spaced practice encourages active processing of information instead of rote learning, often associated with massed practice (Underwood 1961; Reynolds and Glaser 1964; Postman and Underwood 1973).

2. Assignments that include opportunities for the exploration of future topics are preferable to assignments that do not (Klinger 1973).

Exploratory practice with future topics should be carefully designed to account for a general lack of specific knowledge and skills associated with those topics. For instance, on each of several days prior to encountering "dry mixture" or "blend" problems, students might be asked to answer exploratory questions like the following:

- If you were to mix 3 pounds of candy altogether worth $4 with 5 pounds of candy altogether worth $6, how many pounds of candy would be in the mixture? How much would the total mixture be worth? How much would the mixture be worth per pound?

- If gum worth 75 cents a pack is mixed with gum worth 50 cents a pack, which of the following answers is the most reasonable price per pack for the gum? Explain. (1) 45 cents (2) 65 cents (3) 90 cents

Before working with equations of the form $ax + b = cx + d$, give students opportunities, physically or diagrammatically, to determine the weight of a ball, given that a balance scale with 3 balls and 5 pounds on one side balances with 2 balls and 12 pounds on the other.

Before teaching uniform motion, have students simulate various types of motion (same direction, opposite direction, round trip) on a number line to answer questions such as these:

- John and Mary are 15 miles apart and walking toward each other. If John is walking at 3 mph and Mary is walking at 2 mph, how long after they start walking will they meet? Where will they meet?

• John begins walking at 3 mph from point A to point B. Four hours later, walking at 5 mph, Mary starts out from point A along the same path as John. When and where will Mary catch up with John? How does the distance from A to B affect your answer?

Exploratory exercises should serve to activate relevant and meaningful information that the learner already has. The learner's store of previously acquired information can be a tremendous influence on his or her ability to understand new concepts and skills. A great deal of information has been assembled that establishes the importance of activated prior knowledge in assisting a learner to learn, comprehend, and remember (Gagné and Briggs 1979; Bransford 1979).

3. *Same-context practice facilitates initial learning; varied-content practice facilitates transfer* (Nitsch, cited in Bransford 1979; DiVesta and Peverly 1984).

If homework exercises from each lesson are distributed over three or four days, each homework assignment will, by design, include quite a variety of exercises. What results from such mixed practice is generally twofold. First, related ideas and skills become cross-referenced in the learner's cognitive structure instead of being stored in relative isolation. Second, learned concepts are progressively refined as they are modified over time (Ausubel 1963).

Initial learning can be facilitated through focused, massed, guided practice on exercises related to the new topic. Such guided practice of initial learning should emphasize the sameness of a collection of new problem types. For instance, during a lesson on factoring the difference of two squares, ample guided practice during class should be given just on factoring the difference of two squares. Later, in the homework assignment, students can be given the opportunity to encounter this type of factoring in a varied context with, for example, factoring out a greatest common factor and, more generally, factoring trinomials into the product of two binomials. The mixed practice encourages students to extract, for example, the fact that factoring trinomials of the form $ax^2 + bx + c$, no matter how complicated, involves finding two binomials, which when multiplied out by the FOIL (First, Outer, Inner, Last) method, yield the given trinomial. With enough varied practice in factoring polynomials, students should be able to transfer the skill to factoring considerably more complex polynomials, such as $(a + 2)^2 - 3(a + 2) - 10$ or $(x - 4)^2 - 9$.

4. *A combination of distributed and exploratory practice is preferable to massed practice* (Holdan 1986).

Certainly, one would expect the benefits of both distributed and exploratory exercises to be better than either alone. Each assignment simultaneously provides for review, reinforcement, and the opportunity to explore future topics, along with, of course, practice with exercises related to new learning. The varied-practice context embodied in the homework assignment encourages active learning of algebra as a body of integrated principles instead of rote learning of seemingly unrelated types of algebra problems.

5. *Different instructional methods used in teaching may lead to structurally different learning outcomes in terms of the quality of transfer* (Mayer and Greeno 1972, 1975; DiVesta and Peverly 1984).

On the one hand, instruction that emphasizes general concepts enhances far transfer, or broad generalization, of what was learned to problems quite unlike those practiced. For instance, in lessons on uniform motion, students receiving instruction that emphasizes *distance as a function of rate and time* should be able to transfer the idea to analogous rate-of-work problems. On the other hand, instruction that emphasizes solution with a formula enhances near transfer, or limited generalization, of what was learned to problems very much like those practiced. For instance, instruction on uniform motion that focuses on always filling in a $d = rt$ chart and relating two of the squares in the chart will enable students to solve those uniform motion problems that can be solved that way, and very well!

Clearly, the quality of instruction and of the guided practice that precedes the homework assignment affects a student's success with homework. For instance, when teaching exponents, emphasize their counting nature. That is, instead of encouraging students to memorize and arbitrarily apply rules such as $(a^x)^y = a^{xy}$ and $a^x a^y = a^{x+y}$, encourage them to apply simple counting principles to determine the number of times x and y are used as factors in expressions such as $(x^3 y^4)^5$.

Instead of arbitrarily teaching the absolute value rule ($|x| = x$ if $x > 0$, x if $x = 0$, $-x$ if $x < 0$), emphasize during instruction that absolute value represents distance from the origin on a number line. For $|x - 3| = 2$, for instance, encourage students to determine first where $x - 3$ could be on a number line if $x - 3$ is more than two units from the origin and then where x would be. Emphasis on the general concepts in instruction and guided practice instead of on the mere application of seemingly arbitrary rules can enhance the problem-solving efforts of your students.

CONCLUSION

If we assume that transfer is a desirable outcome of mathematics instruction, then homework that combines both distributed and exploratory practice appears to be the way to go. The skills and concepts of mathematics are not taught in isolation from one another; rather, the strongly prerequisite nature of mathematics lends itself nicely to a combination of distributed and exploratory practice. For the most part, single lessons have prerequisite concepts and skills developed in previous lessons. Furthermore, any single lesson is taught with the expectation that its skills and concepts will be needed for learning and understanding in future lessons. The combination of distributed and exploratory practice provides spaced review and reinforcement; it activates a student's prior understandings so that new learning can be meaningfully attached or assimilated; it builds into instruction same- and varied-context practice that facilitates both initial learning and later application.

The principles of effective practice offered and discussed above are not prescriptive; they should serve as guides to making instructionally sound decisions regarding homework. In any instance, homework should be assigned only after students are ready to engage in it. Clear, direct instruction with adequate guided, massed practice and feedback needs to occur before any assignment that requires independent retrieval and response formation. Then, and only then, should students be expected to engage in independent practice of the kind suggested here.

REFERENCES

Ausubel, David P. *The Psychology of Meaningful Verbal Learning*. New York: Grune & Stratton, 1963.

Bransford, John D. *Human Cognition: Learning, Understanding and Remembering*. Belmont, Calif.: Wadsworth, 1979.

Butcher, John E. "Comparison of the Effects of Distributed and Massed Problem Assignments on the Homework of Ninth-Grade Algebra Students." (Ph.D. diss., Rutgers University, 1975.) *Dissertation Abstracts International* 36 (1976):6586A. (University Microfilms No. 76-8683)

Coulter, Frank. "Homework: A Neglected Research Area." *British Educational Research Journal* 5(1) (1979): 21–33.

DiVesta, Francis J., and Stephen T. Peverly. "The Effects of Encoding Variability, Processing Activity, and Rule-Example Sequence on the Transfer of Conceptual Rules." *Journal of Educational Psychology* 76 (February 1984): 108–19.

Dodes, Irving A. "Planned Instruction." In *The Learning of Mathematics: Its Theory and Practice*, Twenty-first Yearbook of the National Council of Teachers of Mathematics, edited by Howard F. Fehr. Washington, D.C.: The Council, 1953.

Frederick, Wayne C., and Herbert J. Walberg. "Learning as a Function of Time." *Journal of Educational Research* 73 (March-April 1980): 183–204.

Gagné, Robert M., and Leslie J. Briggs. *Principles of Instructional Design*. 2d ed. New York: Holt, Rinehart & Winston, 1979.

Goldstein, Avram. "Does Homework Help? A Review of Research." *Elementary School Journal* 1 (January 1960): 212–14.

Holdan, Edmund G. "A Comparison of the Effects of Traditional, Exploratory, Distributed, and a Combination of Distributed and Exploratory Practice on Initial Learning, Transfer, and Retention of Verbal Problem Types in First-Year Algebra." (Ph.D. diss., Pennsylvania State University, 1985.) *Dissertation Abstracts International* 46 (1986): 2542A.

Keith, Timothy Z. "Time Spent on Homework and High School Grades: A Large-Sample Path Analysis." *Journal of Educational Psychology* 74 (April 1982): 248–53.

Klinger, William R. "Effect of Distribution of Earlier Concepts as Preliminary Exercises upon Achievement in a Remedial Mathematics Course at the College Level." (Ph.D. diss., Ohio State University, 1973.) *Dissertation Abstracts International* 34 (1974): 4567–68A. (University Microfilms No. 74-3220)

Mayer, Richard E., and James G. Greeno. "Structural Differences between Learning Outcomes Produced by Different Instructional Methods." *Journal of Educational Psychology* 63 (April 1972): 165–73.

————. "Effects of Meaningfulness and Organization on Problem Solving and Computability Judgments." *Memory and Cognition* 3 (July 1975): 356–62.

Postman, Leo, and Benton J. Underwood. "Critical Issues in Interference Theory." *Memory and Cognition* 1 (January 1973): 19–40.

Reynolds, James H., and Robert Glaser. "Effects of Repetition and Spaced Review upon Retention of a Complex Learning Task." *Journal of Educational Psychology* 55 (October 1964): 297–308.

Rickards, John P. "Homework." In *Encyclopedia of Educational Research*, vol. 2, 5th ed., edited by Howard E. Mitzel, John H. Best, and William Rabinowitz, pp. 831–34. New York: Macmillan, Free Press, 1982.

Underwood, Benton J. "Ten Years of Massed Practice on Distributed Practice." *Psychological Review* 68 (July 1961): 229–47.

CAN YOUR ALGEBRA CLASS SOLVE THIS?

Problem 23. Find two fractions evenly spaced between these two:

$$\frac{3}{a} \text{ and } \frac{7}{2a}$$

Solution on page 248

CAN YOUR ALGEBRA CLASS SOLVE THIS?

Problem 24. Three men form a corporation. Mr. A invests x dollars, Mr. B invests y dollars, and Mr. C invests y dollars. What part of a $100 profit should Mr. A receive if each man shares in the profits in proportion to the amount he invested?

Solution on page 248

SOLUTIONS to "Algebra Problems for Classroom Use"

(Problems are interspersed among the chapters of the book.)

1. $a = -6$

2. $c = 16$

3. $x \in \{1, 2, 3, 4, 5\}$

4. $\dfrac{100dp}{c}$

5. 4 (*Hint:* $\dfrac{1}{x} + \dfrac{1}{y} = \dfrac{y + x}{xy}$)

6. -20.5

7. 1

8. $x^2 = 80$ (*Hint:* Cube, then look for a quadratic.)

9. $x + y = 27$

10. $k = 9$ or $k = -7$

11. equal and real

12. sum $= 5s + 80n$

13. $\left(\dfrac{3}{2}, \dfrac{1}{2}\right)$

14. $x = 4a$

15. $A = 3; B = -1$

16. $x = \pm\sqrt{2}$

17. $x = -1$

18. $a + b + c = 0$ ($a = 1, b = -6, c = 5$)

19. $x + y = \pm 4$

20. $\$40$

21. 16

22. $\dfrac{y}{x} \times 100$

23. $\dfrac{19}{6a}$ and $\dfrac{20}{6a}$

24. $\dfrac{x}{x + xy} \cdot 100$ dollars

248